P9-CKN-522

LAWS OF FORM

BY THE SAME AUTHOR

Probability and scientific inference

Only two can play this game (published as by James Keys)

LAWS OF FORM

G SPENCER BROWN

THE JULIAN PRESS, INC. *Publishers*
NEW YORK

First published in U.S.A. by
The Julian Press, Inc.
150 Fifth Avenue
New York, N.Y. 10011
1972

First published in Great Britain by
George Allen and Unwin Ltd.
London, 1969

Library of Congress Catalog Card Number: 72-80668

Manufactured in the United States of America

A NOTE ON THE MATHEMATICAL APPROACH

The theme of this book is that a universe comes into being when a space is severed or taken apart. The skin of a living organism cuts off an outside from an inside. So does the circumference of a circle in a plane. By tracing the way we represent such a severance, we can begin to reconstruct, with an accuracy and coverage that appear almost uncanny, the basic forms underlying linguistic, mathematical, physical, and biological science, and can begin to see how the familiar laws of our own experience follow inexorably from the original act of severance. The act is itself already remembered, even if unconsciously, as our first attempt to distinguish different things in a world where, in the first place, the boundaries can be drawn anywhere we please. At this stage the universe cannot be distinguished from how we act upon it, and the world may seem like shifting sand beneath our feet.

Although all forms, and thus all universes, are possible, and any particular form is mutable, it becomes evident that the laws relating such forms are the same in any universe. It is this sameness, the idea that we can find a reality which is independent of how the universe actually appears, that lends such fascination to the study of mathematics. That mathematics, in common with other art forms, can lead us beyond ordinary existence, and can show us something of the structure in which all creation hangs together, is no new idea. But mathematical texts generally begin the story somewhere in the middle, leaving the reader to pick up the thread as best he can. Here the story is traced from the beginning.

Unlike more superficial forms of expertise, mathematics is a way of saying less and less about more and more. A mathematical text is thus not an end in itself, but a key to a world beyond the compass of ordinary description.

An initial exploration of such a world is usually undertaken in the company of an experienced guide. To undertake it alone, although possible, is perhaps as difficult as to enter the world of music by attempting, without personal guidance, to read the score-sheets of a master composer, or to set out on a first solo flight in an aeroplane with no other preparation than a study of the pilots' manual.

Although the notes at the end of the text may to some extent make up for, they cannot effectively replace, such personal guidance. They are designed to be read in conjunction with the text, and it may in fact be helpful to read them first.

The reader who is already familiar with logic, in either its traditional or its symbolic form, may do well to begin with Appendix 2, referring through the Index of Forms to the text whenever necessary.

Tho' obscur'd, this is the form of the Angelic land.

William Blake: *America*

PREFACE TO THE FIRST AMERICAN EDITION

Apart from the standard university logic problems, which the calculus published in this text renders so easy that we need not trouble ourselves further with them, perhaps the most significant thing, from the mathematical angle, that it enables us to do is to use complex values in the algebra of logic. They are the analogs, in ordinary algebra, to complex numbers $a + b\sqrt{-1}$. My brother and I had been using their Boolean counterparts in practical engineering for several years before realizing what they were. Of course, being what they are, they work perfectly well, but understandably we felt a bit guilty about using them, just as the first mathematicians to use 'square roots of negative numbers' had felt guilty, because they too could see no plausible way of giving them a respectable academic meaning. All the same, we were quite sure there was a perfectly good theory that would support them, if only we could think of it.

The position is simply this. In ordinary algebra, complex values are accepted as a matter of course, and the more advanced techniques would be impossible without them. In Boolean algebra (and thus, for example, in all our reasoning processes) we disallow them. Whitehead and Russell introduced a special rule, which they called the Theory of Types, expressly to do so. Mistakenly, as it now turns out. So, in this field, the more advanced techniques, although not impossible, simply don't yet exist. At the present moment we are constrained, in our reasoning processes, to do it the way it was done in Aristotle's day. The poet Blake might have had some insight into this, for in 1788 he wrote that 'reason, or the ratio of all we have already known, is not the same that it shall be when we know more.'

Recalling Russell's connexion with the Theory of Types, it

was with some trepidation that I approached him in 1967 with the proof that it was unnecessary. To my relief he was delighted. The Theory was, he said, the most arbitrary thing he and Whitehead had ever had to do, not really a theory but a stopgap, and he was glad to have lived long enough to see the matter resolved.

Put as simply as I can make it, the resolution is as follows. All we have to show is that the self-referential paradoxes, discarded with the Theory of Types, are no worse than similar self-referential paradoxes, which are considered quite acceptable, in the ordinary theory of equations.

The most famous such paradox in logic is in the statement, 'This statement is false.'

Suppose we assume that a statement falls into one of three categories, true, false, or meaningless, and that a meaningful statement that is not true must be false, and one that is not false must be true. The statement under consideration does not appear to be meaningless (some philosophers have claimed that it is, but it is easy to refute this), so it must be true or false. If it is true, it must be, as it says, false. But if it is false, since this is what it says, it must be true.

It has not hitherto been noticed that we have an equally vicious paradox in ordinary equation theory, because we have carefully guarded ourselves against expressing it this way. Let us now do so.

We will make assumptions analogous to those above. We assume that a number can be either positive, negative, or zero. We assume further that a nonzero number that is not positive must be negative, and one that is not negative must be positive. We now consider the equation

$$x^2 + 1 = 0.$$

Transposing, we have

$$x^2 = -1,$$

and dividing both sides by x gives

$$x = \frac{-1}{x}.$$

We can see that this (like the analogous statement in logic) is self-referential: the root-value of x that we seek must be put back into the expression from which we seek it.

Mere inspection shows us that x must be a form of unity, or the equation would not balance numerically. We have assumed only two forms of unity, $+1$ and -1, so we may now try them each in turn. Set $x = +1$. This gives

$$+1 = \frac{-1}{+1} = -1$$

which is clearly paradoxical. So set $x = -1$. This time we have

$$-1 = \frac{-1}{-1} = +1$$

and it is equally paradoxical.

Of course, as everybody knows, the paradox in this case is resolved by introducing a fourth class of number, called *imaginary,* so that we can say the roots of the equation above are $\pm i$, where i is a new kind of unity that consists of a square root of minus one.

What we do in Chapter 11 is extend the concept to Boolean algebras, which means that a valid argument may contain not just three classes of statement, but four: true, false, meaningless, and imaginary. The implications of this, in the fields of logic, philosophy, mathematics, and even physics, are profound.

What is fascinating about the imaginary Boolean values, once we admit them, is the light they apparently shed on our concepts of matter and time. It is, I guess, in the nature of us all to wonder why the universe appears just the way it does. Why, for example, does it not appear more symmetrical? Well, if you will be kind enough, and patient enough, to bear with me through the argument as it develops itself in this text, you will I think see, even though we begin it as symmetrically as we know how, that it becomes, of its own accord, less and less so as we proceed.

G SPENCER BROWN

Cambridge, England
Maundy Thursday 1972

PREFACE

The exploration on which this work rests was begun towards the end of 1959. The subsequent record of it owes much, in its early stages, to the friendship and encouragement of Lord Russell, who was one of the few men at the beginning who could see a value in what I proposed to do. It owes equally, at a later stage, to the generous help of Dr J C P Miller, Fellow of University College and Lecturer in Mathematics in the University of Cambridge, who not only read the successive sets of printer's proofs, but also acted as an ever-available mentor and guide, and made many suggestions to improve the style and accuracy of both text and context.

In 1963 I accepted an invitation of Mr H G Frost, Staff Lecturer in Physical Sciences in the Department of Extra-mural Studies in the University of London, to give a course of lectures on the mathematics of logic. The course was later extended and repeated annually at the Institute of Computer Science in Gordon Square, and from it sprang some of the context in the notes and appendices of this essay. I was also enabled, through the help of successive classes of pupils, to extend and sharpen the text.

Others helped, but cannot, alas, all be mentioned. Of these the publishers (including their readers and their technical artist) were particularly cooperative, as were the printers, and, before this, Mrs Peter Bragg undertook the exacting task of preparing a typescript. Finally I should mention the fact that an original impetus to the work came from Mr I V Idelson, General Manager of Simon-MEL Distribution Engineering, the techniques here recorded being first developed not in respect of questions of logic, but in response to certain unsolved problems in engineering.

Richmond, August 1968

Acknowledgment

The author and publishers acknowledge the kind permission of Mr J Lust, of the University of London School of Oriental and African Studies, to photograph part of a facsimile copy of the 12th century Fukien print of the Tao Tê Ching in the old Palace Museum, Peking.

INTRODUCTION

A principal intention of this essay is to separate what are known as algebras of logic from the subject of logic, and to re-align them with mathematics.

Such algebras, commonly called Boolean, appear mysterious because accounts of their properties at present reveal nothing of any mathematical interest about their arithmetics. Every algebra has an arithmetic, but Boole designed[1] his algebra to fit logic, which is a possible interpretation of it, and certainly not its arithmetic. Later authors have, in this respect, copied Boole, with the result that nobody hitherto appears to have made any sustained attempt to elucidate and to study the primary, non-numerical arithmetic of the algebra in everyday use which now bears Boole's name.

When I first began, some seven years ago, to see that such a study was needed, I thus found myself upon what was, mathematically speaking, untrodden ground. I had to explore it inwards to discover the missing principles. They are of great depth and beauty, as we shall presently see.

In recording this account of them, I have aimed to write so that every special term shall be either defined or made clear by its context. I have assumed on the part of the reader no more than a knowledge of the English language, of counting, and of how numbers are commonly represented. I have allowed myself the liberty of writing somewhat more technically in this introduction and in the notes and appendices which follow the text, but even here, since the subject is of such general interest, I have endeavoured, where possible, to keep the account within the reach of a non-specialist.

Accounts of Boolean algebras have up to now been based on sets of postulates. We may take a postulate to be a statement

[1] George Boole, *The mathematical analysis of logic*, Cambridge, 1847.

which is accepted without evidence, because it belongs to a set of such statements from which it is possible to derive other statements which it happens to be convenient to believe. The chief characteristic which has always marked such statements has been an almost total lack of any spontaneous appearance of truth[2]. Nobody pretends, for example, that Sheffer's equations[3] are mathematically evident, for their evidence is not apparent apart from the usefulness of equations which follow from them. But in the primary arithmetic developed in this essay, the initial equations can be seen to represent two very simple laws of indication which, whatever our views on the nature of their self-evidence, at least recommend themselves to the findings of common sense. I am thus able to present (Appendix 1), apparently for the first time, proofs of each of Sheffer's postulates, and hence of all Boolean postulates, as theorems about an axiomatic system which is seen to rest on the fundamental ground of mathematics.

Working outwards from this fundamental source, the general form of mathematical communication, as we understand it today, tends to grow quite naturally under the hand that writes it. We have a definite system, we name its parts, and we adopt, in many cases, a single symbol to represent each name. In doing this, forms of expression are called inevitably out of the need for them, and the proofs of theorems, which are at first seen to be little more than a relatively informal direction of attention to the complete range of possibilities, become more and more recognizably indirect and formal as we proceed from our original conception. At the half-way point the algebra, in all its representative completeness, is found to have grown imperceptibly out of the arithmetic, so that by the time we have started to work in it we are already fully acquainted with its formalities and possibilities without anywhere having set out with the intention of describing them as such.

One of the merits of this form of presentation is the gradual building up of mathematical notions and common forms of procedure without any apparent break from common sense.

[2] Cf Alfred North Whitehead and Bertrand Russell, *Principia mathematica*, Vol. I, 2nd edition, Cambridge, 1927, p v.

[3] Henry Maurice Sheffer, *Trans. Amer. Math. Soc.*, 14 (1913) 481–8.

The discipline of mathematics is seen to be a way, powerful in comparison with others, of revealing our internal knowledge of the structure of the world, and only by the way associated with our common ability to reason and compute.

Even so, the orderly development of mathematical conventions and formulations stage by stage has not been without its problems on the reverse side. A person with mathematical training, who may automatically use a whole range of techniques without questioning their origin, can find himself in difficulties over an early part of the presentation, in which it has been necessary to develop an idea using only such mathematical tools as have already been identified. In some of these cases we need to derive a concept for which the procedures and techniques already developed are only just adequate. The argument, which is maximally elegant at such a point, may thus be conceptually difficult to follow.

One such case, occurring in Chapter 2, is the derivation of the second of the two primitive equations of the calculus of indications. There seems to be such universal difficulty in following the argument at this point, that I have restated it less elegantly in the notes on this chapter at the end of the text. When this is done, the argument is seen to be so simple as to be almost mathematically trivial. But it must be remembered that, according to the rigorous procedure of the text, no principle may be used until it has been either called into being or justified in terms of other principles already adopted. In this particular instance, we make the argument easy by using ordinary substitution. But at the stage in the essay where it becomes necessary to formulate the second primitive equation, no principle of substitution has yet been called into being, since its use and justification, which we find later in the essay itself, depends in part upon the existence of the very equation we want to establish.

In Appendix 2, I give a brief account of some of the simplifications which can be made through using the primary algebra as an algebra of logic. For example, there are no primitive propositions. This is because we have a basic freedom, not granted to other algebras of logic, of access to the arithmetic whenever we please. Thus each of Whitehead and Russell's five primitive implications [2, pp 96–7] can be equated mathematically with a single constant. The constant, if it were a proposition, would be *the* primitive implication. But in fact, being arithmetical, it cannot represent a proposition.

A point of interest in this connexion is the development of the idea of a variable solely from that of the operative constant. This comes from the fact that the algebra represents our ability to consider the form of an arithmetical equation irrespective of the appearance, or otherwise, of this constant in certain specified places. And since, in the primary arithmetic, we are not presented, apparently, with two kinds of constant, such as 5, 6, etc and $+$, $\times$, etc, but with expressions made up, apparently, of similar constants each with a single property, the conception of a variable comes from considering the irrelevant presence or absence of this property. This lends support to the view, suggested[4] by Wittgenstein, that variables in the calculus of propositions do not in fact represent the propositions in an expression, but only the truth-functions of these propositions, since the propositions themselves cannot be equated with the mere presence or absence of a given property, while the possibility of their being true or not true can.

Another point of interest is the clear distinction, with the primary algebra and its arithmetic, that can be drawn between the proof of a theorem and the demonstration of a consequence. The concepts of theorem and consequence, and hence of proof and demonstration, are widely confused in current literature, where the words are used interchangeably. This has undoubtedly created spurious difficulties. As will be seen in the statement of the completeness of the primary algebra (theorem 17), what is to be proved becomes strikingly clear when the distinction is properly maintained. (A similar confusion is apparent, especially in the literature of symbolic logic, of the concepts of axiom and postulate.)

It is possible to develop the primary algebra to such an extent that it can be used as a restricted (or even as a full) algebra of numbers. There are several ways of doing this, the most convenient of which I have found is to limit condensation

[4] Ludwig Wittgenstein, *Tractatus logico-philosophicus*, London, 1922, propositions 5 sq.

in the arithmetic, and thus to use a number of crosses in a given space to represent either the corresponding number or its image. When this is done it is possible to see plainly some at least of the evidence for Gödel's and Church's theorems[5,6] of decision. But with the rehabilitation of the paradoxical equations undertaken in Chapter 11, the meaning and application of these theorems now stands in need of review. They certainly appear less destructive than was hitherto supposed.

I aimed in the text to carry the development only so far as to be able to consider reasonably fully all the forms that emerge at each stage. Although I indicate the expansion into complex forms in Chapter 11, I otherwise try to limit the development so as to render the account, as far as it goes, complete.

Most of the theorems are original, at least as theorems, and their proofs therefore new. But some of the later algebraic and mixed theorems, occurring in what is at this stage familiar ground, are already known and have, in other forms, been proved before. In all of these cases I have been able to find what seem to be clearer, simpler, or more direct proofs, and in most cases the theorems I prove are more general. For example, the nearest approach to my theorem 16 seems to be a weaker and less central theorem apparently first proved[7] by Quine, as a lemma to a completeness proof for a propositional calculus. It was only after contemplating this theorem for some two years that I found the beautiful key by which it is seen to be true for all possible algebras, Boolean or otherwise.

In arriving at proofs, I have often been struck by the apparent alignment of mathematics with psycho-analytic theory. In each discipline we attempt to find out, by a mixture of contemplation, symbolic representation, communion, and communication, what it is we already know. In mathematics, as in other forms of self-analysis, we do not have to go exploring the physical world to find what we are looking for. Any child of ten, who can multiply and divide, already knows, for example,

[5] Kurt Gödel, *Monatshefte für Mathematik und Physik*, 38 (1931) 172–98.

[6] Alonzo Church, *J. Symbolic Logic*, 1 (1936) 40–1, 101–2.

[7] W V Quine, *J. Symbolic Logic*, 3 (1938) 37–40.

that the sequence of prime numbers is endless. But if he is not shown Euclid's proof, it is unlikely that he will ever find out, before he dies, that he knows.

This analogy suggests that we have a direct awareness of mathematical form as an archetypal structure. I try in the final chapter to illustrate the nature of this awareness. In any case, questions of pure probability alone would lead us to suppose that some degree of direct awareness is present throughout mathematics.

We may take it that the number of statements which might or might not be provable is unlimited, and it is evident that, in any large enough finite sample, untrue statements, of those bearing any useful degree of significance, heavily outnumber true statements. Thus in principle, if there were no innate sense of rightness, a mathematician would attempt to prove more false statements than true ones. But in practice he seldom attempts to prove any statement unless he is already convinced of its truth. And since he has not yet proved it, his conviction must arise, in the first place, from considerations other than proof.

Thus the codification of a proof procedure, or of any other directive process, although at first useful, can later stand as a threat to further progress. For example, we may consider the largely unconscious, but now codified, limitation of the reasoning (as distinct from the computative) parts of proof structures to the solution of Boolean equations of the first degree. As we see in Chapter 11, and in the notes thereto, the solution of equations of higher degree is not only possible, but has been undertaken by switching engineers on an *ad hoc* basis for some half a century or more. Such equations have hitherto been excluded from the subject matter of ordinary logic by the Whitehead-Russell theory of types [2, pp 37 sq, e.g. p 77].

I show in the text that we can construct an implicit function of itself so that it re-enters its own space at either an odd or an even depth. In the former case we find the possibility of a self-denying equation of the kind these authors describe. In such a case, the roots of the equation so set up are imaginary. But in the latter case we find a self-confirming equation which is satisfied, for some given configuration of the variables, by two real roots.

I am able, by this consideration, to rehabilitate[8] the formal structure hitherto discarded with the theory of types. As we now see, the structure can be identified in the more general theory of equations, behind which there already exists a weight of mathematical experience.

One prospect of such a rehabilitation, which could repay further attention, comes from the fact that, although Boolean equations of the first degree can be fully represented on a plane surface, those of the second degree cannot be so represented. In general, an equation of degree k requires, for its representation, a surface of genus $k - 1$. D J Spencer Brown and I found evidence, in unpublished work undertaken in 1962–5, suggesting that both the four-colour theorem and Goldbach's theorem are undecidable with a proof structure confined to Boolean equations of the first degree, but decidable if we are prepared to avail ourselves of equations of higher degree.

One of the motives prompting the furtherance of the present work was the hope of bringing together the investigations of the inner structure of our knowledge of the universe, as expressed in the mathematical sciences, and the investigations of its outer structure, as expressed in the physical sciences. Here the work of Einstein, Schrödinger, and others seems to have led to the realization of an ultimate boundary of physical knowledge in the form of the media through which we perceive it. It becomes apparent that if certain facts about our common experience of perception, or what we might call the inside world, can be revealed by an extended study of what we call, in contrast, the outside world, then an equally extended study of this inside world will reveal, in turn, the facts first met with in the world outside: for what we approach, in either case, from one side or the other, is the common boundary between them.

I do not pretend to have carried these revelations very far,

[8] For a history of the earlier essays to rehabilitate, on a logical rather than on a mathematical basis, something of what was discarded, see Abraham A Fraenkel and Yehoshua Bar-Hillel, *Foundations of set theory*, Amsterdam, 1958, pp 136–95.

or that others, better equipped, could not carry them further. I hope they will. My conscious intention in writing this essay was the elucidation of an indicative calculus, and its latent potential, becoming manifest only when the realization of this intention was already well advanced, took me by surprise.

I break off the account at the point where, as we enter the third dimension of representation with equations of degree higher than unity, the connexion with the basic ideas of the physical world begins to come more strongly into view. I had intended, before I began writing, to leave it here, since the latent forms that emerge at this, the fourth departure from the primary form (or the fifth departure, if we count from the void) are so many and so varied that I could not hope to present them all, even cursorily, in one book.

Medawar observes[9] that the standard form of presentation required of an ordinary scientific paper represents the very reverse of what the investigator was in fact doing. In reality, says Medawar, the hypothesis is first posited, and becomes the medium through which certain otherwise obscure facts, later to be collected in support of it, are first clearly seen. But the account in the paper is expected to give the impression that such facts first suggested the hypothesis, irrespective of whether this impression is truly representative.

In mathematics we see this process in reverse. The mathematician, more frequently than he is generally allowed to admit, proceeds by experiment, inventing and trying out hypotheses to see if they fit the facts of reasoning and computation with which he is presented. When he has found a hypothesis which fits, he is expected to publish an account of the work in the reverse order, so as to deduce the facts from the hypothesis.

I would not recommend that we should do otherwise, in either field. By all accounts, to tell the story backwards is convenient and saves time. But to pretend that the story was actually lived backwards can be extremely mystifying.

In view of this apparent reversal, Laing suggests[10] that what

[9] P B Medawar, Is the Scientific Paper a Fraud, *The Listener*, 12th September 1963, pp 377–8.

[10] R D Laing, *The politics of experience and the bird of paradise*, London, 1967, pp 52 sq.

in empirical science are called *data*, being in a real sense *arbitrarily* chosen by the nature of the hypothesis already formed, could more honestly be called *capta*. By reverse analogy, the facts of mathematical science, appearing at first to be arbitrarily chosen, and thus *capta*, are not really arbitrary at all, but absolutely determined by the nature and coherence of our being. In this view we might consider the facts of mathematics to be the real *data* of experience, for only these appear to be, in the final analysis, inescapable.

Although I have undertaken, to the best of my ability, to preserve, in the text itself, what is thus inescapable, and thereby timeless, and otherwise to discard what is temporal, I am under no illusion of having entirely succeeded on either count. That one can *not*, in such an undertaking, succeed perfectly, seems to me to reside in the manifest imperfection of the state of *particular existence*, in any form at all. (Cf Appendix 2.) The work of any human author must be to some extent idiosyncratic, even though he may know his personal ego to be but a fashionable garb to suit the mode of the present rather than the mean of past and future in which his work will come to rest. To this extent, mode or fashion is inevitable at the expense of mean or meaning, or there can be no connexion of what is peripheral, and has to be regarded, with what is central, and has to be divined.

A major aspect of the language of mathematics is the degree of its formality. Although it is true that we are concerned, in mathematics, to provide a shorthand for what is actually said, this is only half the story. What we aim to do, in addition, is to provide a more general form in which the ordinary language of experience is seen to rest. As long as we confine ourselves to the subject at hand, without extending our consideration to what it has in common with other subjects, we are not availing ourselves of a truly mathematical mode of presentation.

What is encompassed, in mathematics, is a transcedence from a given state of vision to a new, and hitherto unapparent, vision beyond it. When the present existence has ceased to make sense, it can still come to sense again through the realization of its form.

Thus the subject matter of logic, however symbolically treated, is not, in as far as it confines itself to the ground of logic, a mathematical study. It becomes so only when we are able to perceive its ground as a part of a more general form, in a process without end. Its mathematical treatment is a treatment of the form in which our way of talking about our ordinary living experience can be seen to be cradled. It is the laws of this form, rather than those of logic, that I have attempted to record.

In making the attempt, I found it easier to acquire an access to the laws themselves than to determine a satisfactory way of communicating them. In general, the more universal the law, the more it seems to resist expression in any particular mode.

Some of the difficulties apparent in reading, as well as in writing, the earlier part of the text come from the fact that, from Chapter 5 backwards, we are extending the analysis through and beyond the point of simplicity where language ceases to act normally as a currency for communication. The point at which this break from normal usage occurs is in fact the point where algebras are ordinarily taken to begin. To extend them back beyond this point demands a considerable unlearning of the current descriptive superstructure which, until it is unlearned, can be mistaken for the reality.

The fact that, in a book, we have to use words and other symbols in an attempt to express what the use of words and other symbols has hitherto obscured, tends to make demands of an extraordinary nature on both writer and reader, and I am conscious, on my side, of how imperfectly I succeed in rising to them. But at least, in the process of undertaking the task, I have become aware (as Boole himself became aware) that what I am trying to say has nothing to do with me, or anyone else, at the personal level. It, as it were, records itself and, whatever the faults in the record, *that which* is so recorded is not a matter of opinion. The only credit I feel entitled to accept in respect of it is for the instrumental labour of making a record which may, if God so disposes, be articulate and coherent enough to be understood in its temporal context.

London, August 1967

無名天地之始

1

THE FORM

We take as given the idea of distinction and the idea of indication, and that we cannot make an indication without drawing a distinction. We take, therefore, the form of distinction for the form.

Definition

Distinction is perfect continence.

That is to say, a distinction is drawn by arranging a boundary with separate sides so that a point on one side cannot reach the other side without crossing the boundary. For example, in a plane space a circle draws a distinction.

Once a distinction is drawn, the spaces, states, or contents on each side of the boundary, being distinct, can be indicated.

There can be no distinction without motive, and there can be no motive unless contents are seen to differ in value.

If a content is of value, a name can be taken to indicate this value.

Thus the calling of the name can be identified with the value of the content.

Axiom 1. The law of calling

The value of a call made again is the value of the call.

That is to say, if a name is called and then is called again, the value indicated by the two calls taken together is the value indicated by one of them.

That is to say, for any name, to recall is to call.

Equally, if the content is of value, a motive or an intention or instruction to cross the boundary into the content can be taken to indicate this value.

Thus, also, the crossing of the boundary can be identified with the value of the content.

Axiom 2. The law of crossing

The value of a crossing made again is not the value of the crossing.

That is to say, if it is intended to cross a boundary and then it is intended to cross it again, the value indicated by the two intentions taken together is the value indicated by none of them.

That is to say, for any boundary, to recross is not to cross.

2

FORMS TAKEN OUT OF THE FORM

Construction

Draw a distinction.

Content

Call it the first distinction.

Call the space in which it is drawn the space severed or cloven by the distinction.

Call the parts of the space shaped by the severance or cleft the sides of the distinction or, alternatively, the spaces, states, or contents distinguished by the distinction.

Intent

Let any mark, token, or sign be taken in any way with or with regard to the distinction as a signal.

Call the use of any signal its intent.

First canon. Convention of intention

Let the intent of a signal be limited to the use allowed to it.

Call this the convention of intention. In general, *what is not allowed is forbidden.*

Knowledge

Let a state distinguished by the distinction be marked with a mark

$$\urcorner$$

of distinction.

Let the state be known by the mark.

Call the state the marked state.

Form

Call the space cloven by any distinction, together with the entire content of the space, the form of the distinction.

Call the form of the first distinction the form.

Name

Let there be a form distinct from the form.

Let the mark of distinction be copied out of the form into such another form.

Call any such copy of the mark a token of the mark.

Let any token of the mark be called as a name of the marked state.

Let the name indicate the state.

Arrangement

Call the form of a number of tokens considered with regard to one another (that is to say, considered in the same form) an arrangement.

Expression

Call any arrangement intended as an indicator an expression.

Value

Call a state indicated by an expression the value of the expression.

Equivalence

Call expressions of the same value equivalent.

Let a sign

$$=$$

of equivalence be written between equivalent expressions.

Now, by axiom 1,

$$\urcorner\ \urcorner = \urcorner\ .$$

Call this the form of condensation.

Instruction

Call the state not marked with the mark the unmarked state.

Let each token of the mark be seen to cleave the space into which it is copied. That is to say, let each token be a distinction in its own form.

Call the concave side of a token its inside.

Let any token be intended as an instruction to cross the boundary of the first distinction.

Let the crossing be from the state indicated on the inside of the token.

Let the crossing be to the state indicated by the token.

Let a space with no token indicate the unmarked state.

Now, by axiom 2,

$$\overline{\overline{\quad}|}| = \qquad .$$

Call this the form of cancellation.

Equation

Call an indication of equivalent expressions an equation.

Primitive equation

Call the form of condensation a primitive equation.

Call the form of cancellation a primitive equation.

Let there be no other primitive equation.

Simple expression

Note that the three forms of arrangement, ⏋⏋, ⏋⏋ (one cross within another), ⏋, and the one absence of form,   , taken from the primitive equations are all, by convention, expressions.

Call any expression consisting of an empty token simple.

Call any expression consisting of an empty space simple.

Let there be no other simple expression.

Operation

We now see that if a state can be indicated by using a token as a name it can be indicated by using the token as an instruction subject to convention. Any token may be taken, therefore, to be an instruction for the operation of an intention, and may itself be given a name

cross

to indicate what the intention is.

Relation

Having decided that the form of every token called cross is to be perfectly continent, we have allowed only one kind of relation between crosses: continence.

Let the intent of this relation be restricted so that a cross is said to contain what is on its inside and not to contain what is not on its inside.

Depth

In an arrangement a standing in a space s, call the number n of crosses that must be crossed to reach a space s_n from s the depth of s_n with regard to s.

Call a space reached by the greatest number of inwards crossings from s a deepest space in a.

Call the space reached by no crossing from s the shallowest space in a.

Thus

$$s_0 = s.$$

Let any cross standing in any space in a cross c be said to be contained in c.

Let any cross standing in the shallowest space in c be said to stand under, or to be covered by, c.

Unwritten cross

Suppose any s_0 to be surrounded by an unwritten cross.

Call the crosses standing under any cross c, written or unwritten, the crosses pervaded by the shallowest space in c.

Pervasive space

Let any given space s_n be said to pervade any arrangement in which s_n is the shallowest space.

Call the space s pervading an arrangement a, whether or not a is the only arrangement pervaded by s, the pervasive space of a.

3

THE CONCEPTION OF CALCULATION

Second canon. Contraction of reference

1 Construct a cross.

2 Mark it with *c*.

3 Let *c* be its name.

4 Let the name indicate the cross.

Let the four injunctions (two of constructive intent, two of conventional intent) above be contracted to the one injunction (of mixed intent) below.

1 Take any cross *c*.

In general, *let injunctions be contracted to any degree in which they can still be followed.*

Third canon. Convention of substitution

In any expression, let any arrangement be changed for an equivalent arrangement.

Step

Call any such change a step.

Let a sign

$$\rightharpoondown$$

stand for the words

is changed to.

Let a barb in the sign indicate the direction of the change.

Direction

A step may now be considered not only with regard to its kind, as in

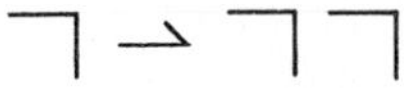

rather than

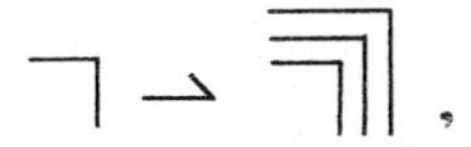

but also with regard to its direction, as in

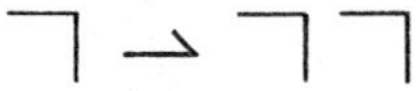

rather than

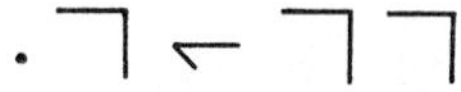

Fourth canon. Hypothesis of simplification

Suppose the value of an arrangement to be the value of a simple expression to which, by taking steps, it can be changed.

Example. To find a value of the arrangement

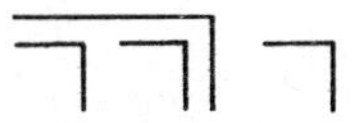

take simplifying steps

condensation

cancellation

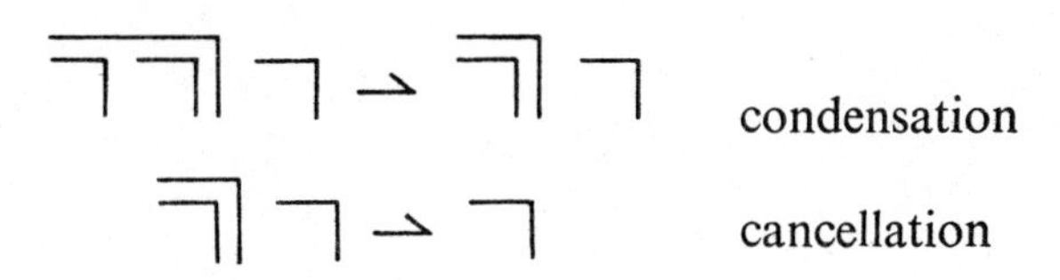

to change it for a simple expression. Now, by the hypothesis of simplification, its value is supposed to be the marked state.

Thus a value for any arrangement can be supposed if the arrangement can be simplified. But it is plain that some arrangements can be simplified in more than one way, and it is conceivable that others might not simplify at all. To show, therefore, that the hypothesis of simplification is a useful determinant of value we shall need to show, at some stage, that any given arrangement will simplify and that all possible procedures for simplifying it will lead to an identical simple expression.

Fifth canon. Expansion of reference

The names hitherto used for the primitive equations suggest steps in the direction of simplicity, and so are not wholly suitable for steps which may in fact be taken in either direction. We therefore expand the form of reference.

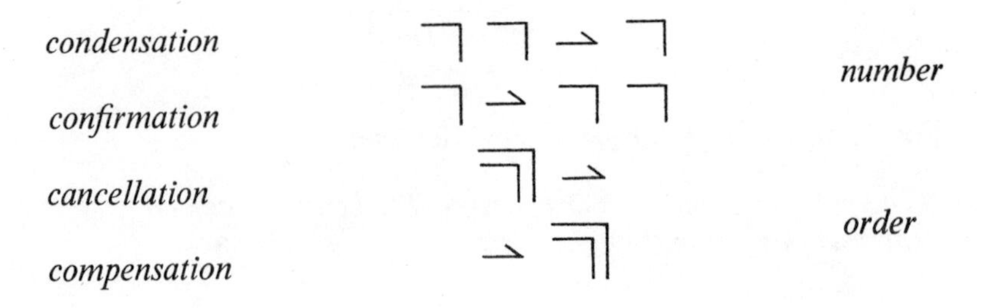

In general, a contraction of reference accompanies an expansion of awareness, and an expansion of reference accompanies a contraction of awareness. If what was done through awareness is to be done by rule, forms of reference must grow (that is to say, divide) to accommodate rules.

Like contraction of reference, of which it is an image, expansion of reference happens, originally, of its own accord. It might at first seem to be a strange procedure, therefore, to call into being a rule permitting it. But we see, if we consider it, that we must call a rule for any process that happens of its own accord, in order to save the convention of intention.

Thus, in general, *let any form of reference be divisible without limit*.

Calculation

Call calculation a procedure by which, as a consequence of steps, a form is changed for another, and call a system of constructions and conventions which allows calculation a calculus.

Initial

The forms of step allowed in a calculus can be defined as all the forms which can be seen in a given set of equations. Call the equations so used to determine these forms the initial equations, or initials, of the calculus.

The calculus of indications

Call the calculus determined by taking the two primitive equations

$$\overline{\quad}|\ \overline{\quad}| = \overline{\quad}| \qquad \text{number}$$

$$\overline{\overline{\quad}|\,}| = \qquad \text{order}$$

as initials the calculus of indications.

Call the calculus limited to the forms generated from direct consequences of these initials the primary arithmetic.

4

THE PRIMARY ARITHMETIC

Initial 1. Number

$$\overline{}| \; \overline{}| \;=\; \overline{}| \qquad \begin{matrix}\text{condense}\\ \rightleftharpoons \\ \text{confirm}\end{matrix}$$

Initial 2. Order

$$\overline{\overline{}|}| \;=\; \qquad \begin{matrix}\text{cancel}\\ \rightleftharpoons \\ \text{compensate}\end{matrix}$$

We shall proceed to distinguish general patterns, called theorems, which can be seen through formal considerations of these initials.

Theorem 1. Form

The form of any finite cardinal number of crosses can be taken as the form of an expression.

That is to say, any conceivable arrangement of any integral number of crosses can be constructed from a simple expression by the initial steps of the calculus.

We may prove this theorem by finding a procedure for simplification: since what can be reduced to a simple expression can, by retracing the steps, be constructed from it.

Proof

Take any such arrangement *a* in a space *s*.

Procedure. Find any deepest space in *a*. It can be found with a finite search since in any given *a* the number of crosses, and thereby the number of spaces, is finite.

Call the space s_d.

Now s_d is either contained in a cross or not contained in a cross.

If s_d is not contained in a cross, then s_d is s and there is no cross in s, and so a is already simple.

If s_d is in a cross c_d, then c_d is empty, since if c_d were not empty s_d would not be deepest.

Now c_d either stands alone in s or does not stand alone in s.

If c_d stands alone in s, then a is already simple.

If c_d does not stand alone in s, then c_d must stand either

(*case 1*) in a space together with another empty cross (if the other cross were not empty s_d would not be deepest) or

(*case 2*) alone in the space under another cross.

Case 1. In this case c_d condenses with the other empty cross. Thereby, one cross is eliminated from a.

Case 2. In this case c_d cancels with the other cross. Thereby, two crosses are eliminated from a.

Now, since each repetition of the procedure used in *case 1* or *case 2* (that is to say, the procedure for an arrangement which is not simple) results in a new arrangement with one or two fewer crosses, there will come a time when, after a finite number of repetitions, a has been either reduced to one cross or eliminated completely.

Thus, in any case, a is simplified.

Therefore, the form of any finite cardinal number of crosses can be taken as the form of an expression.

Theorem 2. Content

If any space pervades an empty cross, the value indicated in the space is the marked state.

Proof

Consider an expression consisting of a part p in a space with an empty cross c_e. It is required to prove that in any case

$$pc_e = c_e.$$

Procedure. Simplify p.

If the procedure reduces p to an empty cross, then the empty cross condenses with c_e and only c_e remains.

If the procedure eliminates p, then only c_e remains.

Thereby, the simplification of every form of pc_e is c_e.

But c_e indicates the marked state.

Therefore, if any space pervades an empty cross, the value indicated in the space is the marked state.

Theorem 3. Agreement

The simplification of an expression is unique.

That is to say, if an expression e simplifies to a simple expression e_s, then e cannot simplify to a simple expression other than e_s.

In simplifying an expression, we may have a choice of steps. Thus the act of simplification cannot be a unique determinant of value unless we can find in it a form independent of this choice.

Now it is clear that, for some expressions, the hypothesis of simplification does provide a unique determinant of value, and we shall proceed to use this fact to show that it provides such a determinant for all expressions.

Let m stand for any number, greater than zero, of such expressions indicating the marked state.

Let n stand for any number of such expressions indicating the unmarked state.

By axiom 1

$$mm = m$$

and

$$nn = n$$

and by simplification or the use of theorem 2,

$$mn = m.$$

Call the value of m a dominant value, and call the value of n a recessive value.

These definitions and considerations may now be summarized in the following rule.

Sixth canon. Rule of dominance

If an expression e in a space s shows a dominant value in s, then the value of e is the marked state. Otherwise, the value of e is the unmarked state.

Also, by definition,

(i) $$m = \overline{\quad}|$$

and

(ii) $$n = $$

so that

$$\overline{m}| = n \qquad \text{(i), cancellation, (ii)}$$

and

$$\overline{n}| = m \qquad \text{(i), (ii).}$$

Proof of theorem 3

Let e stand in the space s_0.

Procedure. Count the number of crossings from s_0 to the deepest space in e. If the number is d, call the deepest space s_d.

By definition, the crosses covering s_d are empty, and they are the only contents of s_{d-1}.

Being empty, each cross in s_{d-1} can be seen to indicate only the marked state, and so the hypothesis of simplification uniquely determines its value.

1 Make a mark m on the outside of each cross in s_{d-1}.

We know by (i) that

$$m = \urcorner .$$

Thus no value in s_{d-1} is changed, since

$\urcorner \rightharpoonup \urcorner\, m$	procedure
$= \urcorner\,\urcorner$	(i)
$= \urcorner$	condensation.

Therefore, the value of e is unchanged.

2 Next consider the crosses in s_{d-2}.

Any cross in s_{d-2} either is empty or covers one or more crosses already marked with m.

If it is empty, mark it with m so that the considerations in *1* apply.

If it covers a mark m, mark it with n.

We know by (ii) that

$$n = \qquad .$$

Thus no value in s_{d-2} is changed.

Therefore, the value of e is unchanged.

3 Consider the crosses in s_{d-3}.

Any cross in s_{d-3} either is empty or covers one or more crosses already marked with *m* or *n*.

If it does not cover a mark *m*, mark it with *m*.

If it covers a mark *m*, mark it with *n*.

In either case, by the considerations in *1* and *2*, no value in s_{d-3} is changed, and so the value of *e* is unchanged.

The procedure in subsequent spaces to s_0 requires no additional consideration.

Thus, by the procedure, each cross in *e* is uniquely marked with *m* or *n*.

Therefore, by the rule of dominance, a unique value of *e* in s_0 is determined.

But the procedure leaves the value of *e* unchanged, and the rules of the procedure are taken from the rules of simplification.

Therefore, the value of *e* determined by the procedure is the same as the value of *e* determined by simplification.

But *e* can be any expression.

Therefore, the simplification of an expression is unique.

Illustration. Let *e* be

The deepest space in *e* is s_4, so mark crosses first in s_3

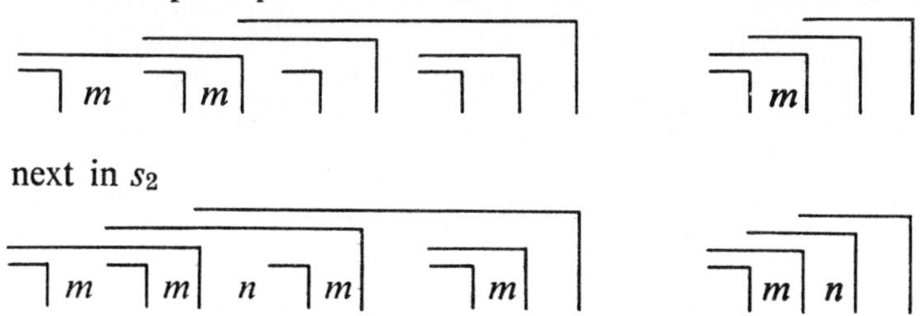

next in s_2

next in s_1

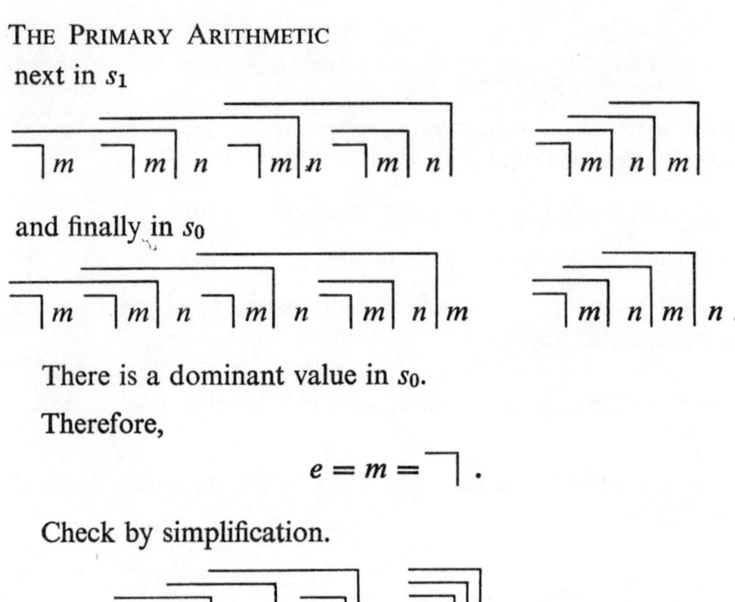

and finally in s_0

There is a dominant value in s_0.

Therefore,

$e = m = \urcorner$.

Check by simplification.

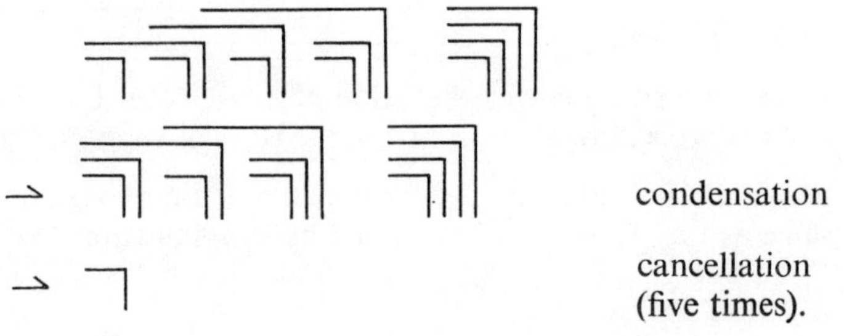

We have shown that the indicators of the two values in the calculus remain distinct when we take steps towards simplicity, thereby justifying the hypothesis of simplification. For completeness we must show that they remain similarly distinct when we take steps away from simplicity.

Theorem 4. Distinction

The value of any expression constructed by taking steps from a given simple expression is distinct from the value of any expression constructed by taking steps from a different simple expression.

Proof

Consider any complex expression e_c constructed as a consequence of steps from a simple expression e_s.

Since each step in the construction of e_c can be retraced, there exists a simplification of e_c which leads to e_s.

But, by theorem 3, all simplifications of e_c agree. Hence all simplifications of e_c lead to e_s.

Thus, by the hypotheses of simplification, the use of which is justified in the proof of theorem 3, the only possible value of e_c is the value of e_s.

But e_s must be one of the simple expressions ⏋ or   , which by definition have distinct values.

Therefore, the value of any expression constructed by taking steps from a given simple expression is distinct from the value of any expression constructed by taking steps from a different simple expression.

Consistency

We have now shown that the two values which the forms of the calculus are intended to indicate are not confused by any step allowed in the calculus and that, therefore, the calculus does in fact carry out its intention.

If, in a calculus intending several indications, they are anywhere confused, then they are everywhere confused, and if they are confused they are not distinguished, and if they are not distinguished they cannot be indicated, and the calculus thereby makes no indication.

A calculus that does not confuse a distinction it intends will be said to be consistent.

A classification of expressions

Expressions of the marked state may be called dominant. The letter *m*, unless otherwise employed, may be taken to indicate a dominant expression.

Expressions of the unmarked state may be called recessive. The letter *n*, unless otherwise employed, may be taken to indicate a recessive expression.

Theorem 5. Identity

Identical expressions express the same value.

In any case,

$$x = x.$$

Proof

By theorems 3 and 4 we see that no step from an expression x can change the value expressed by x.

Therefore, any expression that can be reached by steps from x must have the same value as x.

But an expression identical with x can be reached by taking steps from x and then retracing them.

Thus any expression identical with x must express the same value as x.

Therefore,

$$x = x$$

in any case.

Theorem 6. Value

Expressions of the same value can be identified.

Proof

If x expresses the same value as y, then both x and y will simplify to the same simple expression, call it e_s.

Let $v = e_s$. Thus v will also simplify to e_s, and so v can be reached from either x or y by taking steps to e_s and then retracing the simplification of v.

Thus

$$x = v$$

and

$$y = v.$$

Therefore, by the convention of substitution, both x and y may be changed for an identical expression v in each case.

But x and y can be any equivalent expressions.

Therefore, expressions of the same value can be identified.

Theorem 7. Consequence

Expressions equivalent to an identical expression are equivalent to one another.

In any case, if

$$x = v$$

and

$$y = v,$$

then

$$x = y.$$

Proof

Let e_s be simple, and let $v = e_s$.

Now, since $x = v$ and $y = v$, e_s can be reached by steps from x and by steps from y.

Procedure. Take the steps from x to e_s, and from e_s retrace the steps from y to e_s.

Thus y is reached by steps from x.

Therefore, if

$$x = v$$

and

$$y = v,$$

then

$$x = y$$

in any case.

Theorem 8. Invariance

If successive spaces s_n, s_{n+1}, s_{n+2} *are distinguished by two crosses, and* s_{n+1} *pervades an expression identical with the whole expression in* s_{n+2}, *then the value of the resultant expression in* s_n *is the unmarked state.*

In any case,

$$\overline{\overline{p\,}\rceil\; p\,}\rceil \;=\; \qquad .$$

Proof

Let $p \;=\; \overline{\;\;}\rceil$. In this case

$$\overline{\overline{p\,}\rceil\; p\,}\rceil \;=\; \overline{\overline{\overline{\;\;}\rceil}\rceil\;\overline{\;\;}\rceil}\rceil \qquad \text{substitution}$$

$$= \qquad\qquad \text{order (twice).}$$

Now let $p \;=\;$. In this case

$$\overline{\overline{p\,}\rceil\; p\,}\rceil \;=\; \overline{\overline{\;\;}\rceil}\rceil \qquad \text{substitution}$$

$$= \qquad\qquad \text{order.}$$

There is no other case of p, theorem 1.

There is no other way of substituting any case of p, theorems 5, 6.

Therefore,

$$\overline{\overline{p\,}\rceil\; p\,}\rceil \;=\;$$

in any case.

Theorem 9. Variance

If successive spaces s_n, s_{n+1}, s_{n+2} *are arranged so that* s_n, s_{n+1} *are distinguished by one cross, and* s_{n+1}, s_{n+2} *are distinguished by two crosses* (s_{n+2} being thus in two divisions), *then the whole expression e in* s_n *is equivalent to an expression, similar in other respects to e, in which an identical expression has been taken out of each division of* s_{n+2} *and put into* s_n.

In any case,

$$\overline{\overline{pr|}\;\overline{qr|}\;|} = \overline{\overline{p|}\;\overline{q|}\;|}\;r.$$

Proof

Let $r = \overline{\;\;|}$.

Thus

$$\overline{\overline{pr|}\;\overline{qr|}\;|} = \overline{\overline{p\;\overline{\;\;|}\;|}\;\overline{q\;\overline{\;\;|}\;|}\;|} \qquad \text{substitution}$$

$$= \overline{\overline{\overline{\;\;|}\;|}\;\overline{\overline{\;\;|}\;|}\;|} \qquad \text{theorem 2 (twice)}$$

$$= \overline{\;\;|} \qquad \text{order (twice)}$$

and

$$\overline{\overline{p|}\;\overline{q|}\;|}\;r = \overline{\overline{p|}\;\overline{q|}\;|}\;\overline{\;\;|} \qquad \text{substitution}$$

$$= \overline{\;\;|} \qquad \text{theorem 2.}$$

Therefore, in this case,

$$\overline{\overline{pr|}\;\overline{qr|}\;|} = \overline{\overline{p|}\;\overline{q|}\;|}\;r \qquad \text{theorem 7.}$$

Now let $r =$.

Thus

$$\overline{\overline{pr|}\;\overline{qr|}\;|} = \overline{\overline{p|}\;\overline{q|}\;|} \qquad \text{substitution}$$

and

$$\overline{\overline{p|}\;\overline{q|}\;|}\;r = \overline{\overline{p|}\;\overline{q|}\;|} \qquad \text{substitution.}$$

Therefore, in this case,

$$\overline{\overline{pr|}\;\overline{qr|}\,|} \;=\; \overline{\overline{p|}\;\overline{q|}\,|}\;r \qquad \text{theorem 7.}$$

There is no other case of r, theorem 1.

There is no other way of substituting any case of r, theorems 5, 6.

Therefore,

$$\overline{\overline{pr|}\;\overline{qr|}\,|} \;=\; \overline{\overline{p|}\;\overline{q|}\,|}\;r$$

in any case.

A classification of theorems

The first four theorems contain a statement of completeness and consistency of representation. Their proofs comprise a justification of the use of the primary arithmetic as a system of indicators of the states distinguished by the first distinction. We call them theorems of representation.

The next three theorems justify the use of certain procedural contractions without which subsequent justifications might become intolerably cumbersome. We call them theorems of procedure.

The last two theorems will serve as a gate of entry into a new calculus. We call them theorems of connexion.

The new calculus will itself give rise to further theorems which, when they describe aspects of the new calculus without direct reference to the old, will be called pure algebraic theorems, or theorems of the second order.

In addition we shall find theorems about the two calculi considered together. The bridge theorem and the theorem of completeness are examples, and we may call them mixed theorems.

5

A CALCULUS TAKEN OUT OF THE CALCULUS

Let tokens of variable form

$$a, b, \ldots$$

indicate expressions in the primary arithmetic.

Let their values be unknown except in as far as, by theorem 5,

$$a = a, \quad b = b, \ldots$$

Let tokens of constant form

$$\overline{}\rceil$$

indicate instructions to cross the boundary of the first distinction according to the conventions already called.

Call any token of variable form after its form.

Call any token of constant form

cross.

Let indications used in the description of theorem 8 be taken out of context so that

$$\overline{\overline{p}\rceil\; p}\rceil \;= \qquad .$$

Call this the form of position.

Let indications used in the description of theorem 9 be taken out of context so that

$$\overline{\overline{pr}\;\overline{qr}} \;=\; \overline{\overline{p}\;\overline{q}}\;r.$$

Call this the form of transposition.

Let the forms of position and transposition be taken as the initials of a calculus.

Let the calculus be seen as a calculus for the primary arithmetic.

Call it the primary algebra.

Algebraic calculation

For algebras, two rules are commonly accepted as implicit in the use of the sign =.

Rule 1. Substitution

If $e = f$, and if h is an expression constructed by substituting f for any appearance of e in g, then $g = h$.

Justification. This rule is a restatement of the arithmetical convention of substitution together with an inference from the theorems of representation.

Rule 2. Replacement

If $e = f$, and if every token of a given independent variable expression v in $e = f$ is replaced by an expression w, it not being necessary for v, w to be equivalent or for w to be independent or variable, and if as a result of this procedure e becomes j and f becomes k, then $j = k$.

Justification. This rule derives from the fact, proved with the theorems of connexion, that we can find equivalent expressions, not identical, which, considered arithmetically, are not wholly revealed. In an equation of such expressions each independent variable indicator stands for an expression which, being unknown except in as far as, by theorem 5, its value must be taken to be the same wherever its indicator appears, may be changed at will. Hence its indicator may also be changed at will, provided only that the change is made to every appearance of the indicator.

Indexing

Numbered members of a class of findings will henceforth be indexed by a capital letter denoting the class followed by a figure denoting the number of the member. The classes will be indexed thus.

Consequence	C
Initial of the primary arithmetic	I
Initial of the primary algebra	J
Rule	R
Theorem	T

Certain equations, designated by E, will also be indexed, but the reference in each chapter will be confined to a separate set. Thus E1, etc, in Chapter 9 will not intentionally be the same equations as E1, etc, in Chapter 8.

6

THE PRIMARY ALGEBRA

Initial 1. Position

J1	$\overline{\overline{p}|\; p}| =$	take out ⇌ put in

Initial 2. Transposition

J2	$\overline{\overline{pr}|\;\overline{qr}|}| = \overline{\overline{p}|\;\overline{q}|}|\; r$	collect ⇌ distribute

We shall proceed to distinguish particular patterns, called consequences, which can be found in sequential manipulations of these initials.

Consequence 1. Reflexion

C1	$\overline{\overline{a}|}| = a$	reflect ⇌ reflect

Demonstration

We first find

$$\overline{\overline{a}|}| = \overline{\overline{\overline{a}|}|\;\overline{a}|}|\;\overline{\overline{a}|}|$$

by J1. We use R2 to convert $\overline{\overline{p}|\; p}| =$ to $\overline{\overline{\overline{a}|}|\;\overline{a}|}| =$ by changing every appearance of p

to an appearance of $\overline{a}\rceil$. We next use R1 to change an appearance of to an appearance of $\overline{\overline{\overline{a}\rceil}\rceil\ \overline{a}\rceil}\rceil$ in the space with the original expression $\overline{\overline{a}\rceil}\rceil$, thus finding

$\overline{\overline{a}\rceil}\rceil = \overline{\overline{a}\rceil}\rceil\ \overline{\overline{\overline{a}\rceil}\rceil\ \overline{a}\rceil}\rceil$.

We next find

$$\overline{\overline{\overline{a}\rceil}\rceil\ \overline{a}\rceil}\rceil\ \overline{\overline{a}\rceil}\rceil = \overline{\overline{\overline{\overline{a}\rceil}\rceil\ \overline{a}\rceil}\rceil\ \overline{\overline{\overline{a}\rceil}\rceil\ a}\rceil}\rceil$$

by J2. We make use of the licence allowed in the definition (p 5) of = to convert $\overline{\overline{pr}\rceil\ \overline{qr}\rceil}\rceil = \overline{\overline{p}\rceil\ \overline{q}\rceil}\rceil r$ to $\overline{\overline{p}\rceil\ \overline{q}\rceil}\rceil r = \overline{\overline{pr}\rceil\ \overline{qr}\rceil}\rceil$. We next use the licence allowed in the definition (p 6) of relation to change this to $\overline{\overline{p}\rceil\ \overline{q}\rceil}\rceil r = \overline{\overline{rp}\rceil\ \overline{rq}\rceil}\rceil$. We then use R2 to change every appearance of p in this equation to an appearance of $\overline{a}\rceil$, thus finding $\overline{\overline{\overline{a}\rceil}\rceil\ \overline{q}\rceil}\rceil r = \overline{\overline{r\ \overline{a}\rceil}\rceil\ \overline{rq}\rceil}\rceil$. We use R2 again to change every appearance of q in this equation to an appearance of a, and then again to change every appearance of r to an appearance of $\overline{\overline{a}\rceil}\rceil$, thus finding $\overline{\overline{\overline{a}\rceil}\rceil\ \overline{a}\rceil}\rceil\ \overline{\overline{a}\rceil}\rceil = \overline{\overline{\overline{\overline{a}\rceil}\rceil\ \overline{a}\rceil}\rceil\ \overline{\overline{\overline{a}\rceil}\rceil\ a}\rceil}\rceil$.

We then find

$$\overline{\overline{\overline{\overline{a}\rceil}\rceil\ \overline{a}\rceil}\rceil\ \overline{\overline{\overline{a}\rceil}\rceil\ a}\rceil}\rceil = \overline{\overline{\overline{\overline{a}\rceil}\rceil\ a}\rceil}\rceil$$

by J1. We found $\overline{\overline{\overline{a}\rceil}\rceil\ \overline{a}\rceil}\rceil =$ for the first equation, and we now only need to use R1 to change the appearance of

$\overline{\overline{\overline{a|}|}\;\overline{a|}|}$ in the space with $\overline{\overline{\overline{a|}|}\;a|}$ to an appearance of

in $\overline{\overline{\overline{\overline{a|}|}\;\overline{a|}|}\;\overline{\overline{\overline{a|}|}\;a|}|}$ to find $\overline{\overline{\overline{\overline{a|}|}\;\overline{a|}|}\;\overline{\overline{\overline{a|}|}\;a|}|}$

$= \overline{\overline{\overline{\overline{a|}|}\;a|}|}$.

We then find

$$\overline{\overline{\overline{\overline{a|}|}\;a|}|} = \overline{\overline{\overline{\overline{a|}|}\;a|}\;\overline{\overline{a|}\;a|}|}$$

by J1. We use R2 to convert $\overline{\overline{p|}\;p|} =$ to $\overline{\overline{a|}\;a|} =$ by changing all p to a, and then use R1 to change to $\overline{\overline{a|}\;a|}$ in the space with $\overline{\overline{\overline{a|}|}\;a|}$, thus finding $\overline{\overline{\overline{\overline{a|}|}\;a|}|} = \overline{\overline{\overline{\overline{a|}|}\;a|}\;\overline{\overline{a|}\;a|}|}$.

We then find

$$\overline{\overline{\overline{\overline{a|}|}\;a|}\;\overline{\overline{a|}\;a|}|} = \overline{\overline{\overline{\overline{a|}|}|}\;\overline{\overline{a|}|}|}\;a$$

by J2, using R2 to change all p to $\overline{\overline{a|}|}$, and then all q to $\overline{a|}$, and then all r to a.

And lastly, we find

$$\overline{\overline{\overline{\overline{a|}|}|}\;\overline{\overline{a|}|}|}\;a = a$$

by J1. We find $\overline{\overline{\overline{\overline{a|}|}|}\;\overline{\overline{a|}|}|} =$ by using R2 to change all p to $\overline{\overline{a}}$, and then use R1 to change $\overline{\overline{\overline{a}}\;\overline{\overline{a}}}$ to

in the space with a, thus finding $\overline{\overline{\overline{a}}\;\overline{\overline{a}}}\;a = a$.

This completes a detailed account of each of six steps. We may now use T7 five times to find

$$\overline{\overline{a}} = a$$

and this completes the demonstration.

We repeat this demonstration, and give subsequent demonstrations, with only the key indices to the procedure.

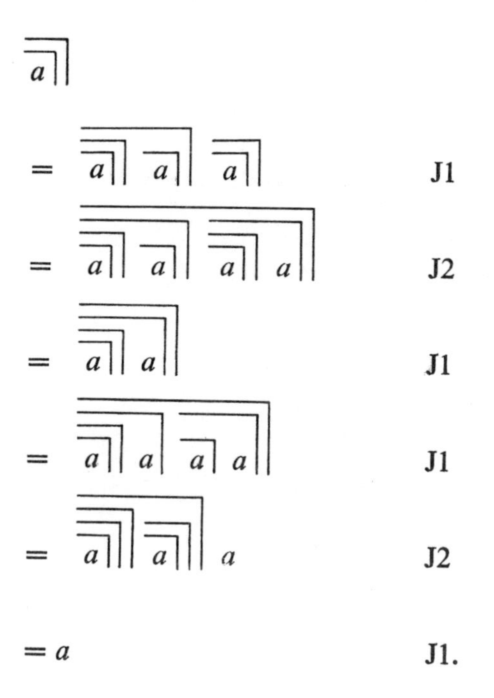

Consequence 2. Generation

C2	$\overline{ab|}\ b = \overline{a|}\ b$	degenerate ⇌ regenerate

Demonstration

$\overline{ab|}\ b$	
$= \overline{\overline{\overline{a|}|}\ b|}\ b$	C1
$= \overline{\overline{\overline{a|}|}\ \overline{\overline{b|}|}|}\ b$	C1
$= \overline{\overline{\overline{a|}\ b|}\ \overline{\overline{b|}\ b|}|}$	J2
$= \overline{\overline{\overline{a|}\ b|}|}$	J1
$= \overline{a|}\ b$	C1.

Consequence 3. Integration

C3	$\overline{\ |}\ a = \overline{\ |}$	reduce ⇌ augment

Demonstration

$\overline{\ |}\ a$	
$= \overline{a|}\ a$	C2
$= \overline{\overline{\overline{a|}\ a|}|}$	C1
$= \overline{\ |}$	J1.

Consequence 4. Occultation

C4 $$\overline{\overline{a\,}|\; b\,}|\; a = a$$ conceal $\rightleftharpoons$ reveal

Demonstration

$$\overline{\overline{a\,}|\; b\,}|\; a$$

$$= \overline{\overline{a\,}|\; ba\,}|\; a \qquad \text{C2}$$

$$= \overline{\overline{ab\,}|\; \;ba\,}|\; a \qquad \text{C2}$$

$$= a \qquad \text{J1.}$$

Consequence 5. Iteration

C5 $$aa = a$$ iterate $\rightleftharpoons$ reiterate

Demonstration

$$aa$$

$$= \overline{\overline{a\,}|\,}|\; a \qquad \text{C1}$$

$$= a \qquad \text{C4.}$$

Consequence 6. Extension

C6 $$\overline{\overline{a\,}|\;\overline{b\,}|\,}|\;\overline{\overline{a\,}|\; b\,}| = a$$ contract $\rightleftharpoons$ expand

Demonstration

$$\overline{\overline{a\,}|\;\overline{b\,}|\,}|\;\overline{\overline{a\,}|\; b\,}|$$

$$= \overline{\overline{\overline{\overline{a\,}|\;\overline{b\,}|\,}|\;\overline{\overline{a\,}|\; b\,}|\,}|\,}| \qquad \text{C1}$$

$$= \overline{\overline{\overline{\overline{b}}\;\overline{b}}\;\overline{a}} \qquad \text{J2}$$

$$= \overline{\overline{a}} \qquad \text{J1}$$

$$= a \qquad \text{C1.}$$

Consequence 7. Echelon

C7 $$\overline{\overline{\overline{a}\;b}\;c} = \overline{ac}\;\overline{\overline{b}\;c}$$ break ⇌ make

Demonstration

$$\overline{\overline{\overline{a}\;b}\;c}$$

$$= \overline{\overline{\overline{a}\;\overline{\overline{b}}}\;c} \qquad \text{C1}$$

$$= \overline{\overline{\overline{ac}\;\overline{\overline{b}\;c}}} \qquad \text{J2}$$

$$= \overline{ac}\;\overline{\overline{b}\;c} \qquad \text{C1.}$$

Consequence 8. Modified transposition

C8 $$\overline{\overline{a}\;\overline{br}\;\overline{cr}} = \overline{\overline{a}\;\overline{b}\;\overline{c}}\;\overline{\overline{a}\;\overline{r}}$$ collect ⇌ distribute

Demonstration

$$\overline{\overline{a}\;\overline{br}\;\overline{cr}}$$

$$= \overline{\overline{a}\;\overline{\overline{\overline{br}\;\overline{cr}}}} \qquad \text{C1}$$

$$= \overline{\overline{a}\;\overline{\overline{\overline{b}\;\overline{c}}\;r}} \qquad \text{J2}$$

$$= \overline{\overline{a}\;\overline{b}\;\overline{c}}\;\overline{\overline{a}\;\overline{r}} \qquad \text{C7.}$$

Consequence 9. Crosstransposition

C9 $\overline{\overline{\overline{b|}\ \overline{r|}|}\ \overline{\overline{a|}\ \overline{r|}|}\ \overline{\overline{x|}\ r|}\ \overline{\overline{y|}\ r|}|}$ $= \overline{r|}\ \overline{ab|}\ \overline{rxy|}$

crosstranspose (collect)

⇌

crosstranspose (distribute)

Demonstration

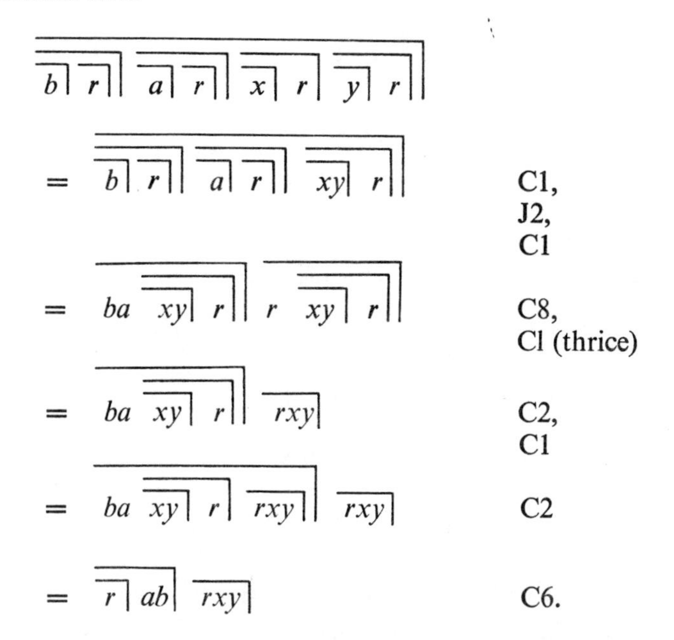

The classification of consequences

In classifying these consequences, there is no need to confine them rigidly to the forms above. The name of a consequence may indicate a part of the consequence as in

$\overline{a|}\ a = \overline{\ \ |}$ integration.

In another case it may include reflexions as in

$$\overline{\overline{a}\rceil\ r}\rceil\ \overline{\overline{b}\rceil\ r}\rceil \;=\; \overline{\overline{ab}\rceil\ r}\rceil \qquad \text{transposition or echelon}$$

and

$$\overline{\overline{b}\rceil\ a}\rceil\ \overline{ba}\rceil \;=\; \overline{a}\rceil \qquad \text{extension.}$$

In yet another case it may indicate a crosstransposed form such as

$$\overline{\overline{a\ b}\rceil\ \overline{a\ \overline{b}\rceil}\rceil}\rceil \;=\; a \qquad \text{extension.}$$

Nor, as we already see in one case, are the classes of consequence properly distinct. What we are doing is to indicate larger and larger numbers of steps in a single indication. This is the dual form of the contraction of a reference, notably the expansion of its content. We shed the labour of calculation by taking a number of steps as one step.

Thus if we consider the equivalence of steps, we find

$$\rightharpoonup\ \rightharpoonup\ =\ \rightharpoonup\ .$$

Also, since to retrace a step can be considered as not to take it, we find

$$\rightharpoonup\ \leftharpoondown\ =\qquad .$$

But now if we allow steps in the indication of steps, we find that the resulting calculus is inconsistent.

Thus

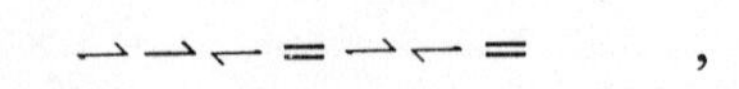

$$\rightharpoonup\ \rightharpoonup\ \leftharpoondown\ =\ \rightharpoonup\ \leftharpoondown\ =\qquad ,$$

or

$$\rightharpoonup\ \rightharpoonup\ \leftharpoondown\ =\ \rightharpoonup\ ,$$

according to which step we take first.

Therefore,

$$= \rightharpoonup ,$$

which suggests that, in any calculation, we regard any number of steps, including zero, as a step.

This agrees with our idea of the nature of a step which, as we have already determined, is not intended to cross a boundary.

A further classification of expressions

The algebraic consideration of the calculus of indications leads to a further distinction between expressions.

Expressions of the marked state may be called integral. The letter *m*, unless otherwise employed, may be taken to indicate an integral expression.

Expressions of the unmarked state may be called disintegral. The letter *n*, unless otherwise employed, may be taken to indicate a disintegral expression.

Expressions of a state consequent on the states of their unknown indicators may be called consequential. The letter *v*, unless otherwise employed, may be taken to indicate a consequential expression.

7

THEOREMS OF THE SECOND ORDER

Theorem 10

The scope of J2 can be extended to any number of divisions of the space s_{n+2}.

In any case,

$$\overline{\overline{a}|\;\overline{b}|\;\ldots}|\; r \;=\; \overline{\overline{ar}|\;\overline{br}|\;\ldots}|\;.$$

Proof

We consider the cases in which s_{n+2} is divided into 0, 1, 2, and more than 2 divisions respectively. In case 0

$$\overline{}|\; r \;=\; \overline{}| \qquad \text{C3.}$$

In case 1

$$\overline{\overline{a}|}|\; r \;=\; ar \qquad \text{C1}$$

$$=\; \overline{\overline{ar}|}| \qquad \text{C1.}$$

In case 2

$$\overline{\overline{b}|\;\overline{a}|}|\; r \;=\; \overline{\overline{br}|\;\overline{ar}|}| \qquad \text{J2.}$$

In case more than 2

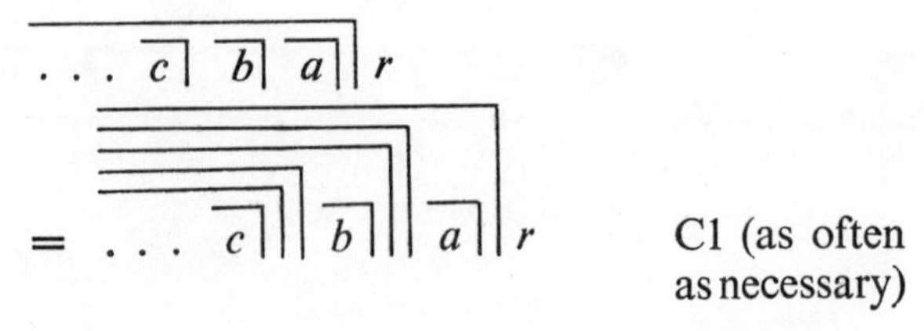

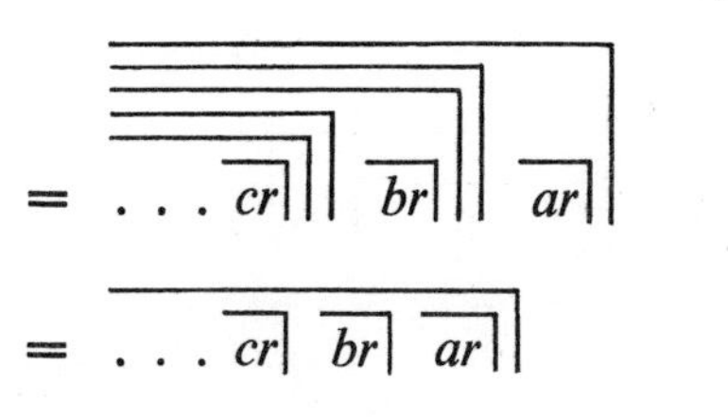

J2 (as often as necessary)

C1 (as often as before).

This completes the proof.

Theorem 11

The scope of C8 can be extended as in T10.

$$\overline{\overline{a}\rceil\ \overline{br}\rceil\ \overline{cr}\rceil \ldots}\rceil = \overline{\overline{a}\rceil\ \overline{b}\rceil\ \overline{c}\rceil \ldots}\rceil\ \overline{a}\rceil\ \overline{r}\rceil\rceil$$

Theorem 12

The scope of C9 can be extended as in T10.

$$\overline{\ldots\ \overline{\overline{b}\rceil\ \overline{r}\rceil}\rceil\ \overline{\overline{a}\rceil\ \overline{r}\rceil}\rceil\ \overline{\overline{x}\rceil\ r}\rceil\ \overline{\overline{y}\rceil\ r}\rceil \ldots}\rceil$$
$$= \overline{\overline{r}\rceil\ ab \ldots}\rceil\ \ \overline{rxy \ldots}\rceil$$

Proofs of T11 and T12 follow from demonstrations as in C8 and C9, using T10 instead of J2.

Theorem 13

The generative process in C2 can be extended to any space not shallower than that in which the generated variable first appears.

Proof

We consider cases in which a variable is generated in spaces 0, 1, and more than 1 space deeper than the space of the variable of origin. In case 0

$$\overline{\overline{\overline{\ldots c}\rceil\ b}\rceil\ a}\rceil\ g = \overline{\overline{\overline{\ldots c}\rceil\ b}\rceil\ a}\rceil\ gg \qquad \text{C5.}$$

In case 1

$$\overline{\overline{\overline{\;.\,.\,.\,c\;}\;b\;}\;a\;}\;g$$

$$= \overline{\overline{\overline{\;.\,.\,.\,c\;}\;b\;}\;ag\;}\;g \qquad \text{C2.}$$

In case more than 1

$$\overline{\overline{\overline{\;.\,.\,.\,c\;}\;b\;}\;a\;}\;g$$

$$= \overline{\overline{\overline{\;.\,.\,.\,c\;}\;b\;}\;ag\;}\;g \qquad \text{C2}$$

$$= \overline{\overline{\overline{\;.\,.\,.\,c\;}\;bg\;}\;ag\;}\;g \qquad \text{C2}$$

$$= \overline{\overline{\overline{\;.\,.\,.\,c\;}\;bg\;}\;a\;}\;g \qquad \text{C2}$$

and so on. Clearly no additional consideration is needed for further generation of g, and it is plain that any space not shallower than that in which g stands can be reached.

It is convenient to consider J2, C2, C8, and C9 as extended by their respective theorems, and to let the name of the initial or consequence denote also the theorem extending it.

Theorem 14. Canon with respect to the constant

From any given expression, an equivalent expression not more than two crosses deep can be derived.

Proof

Suppose that a given expression e has i deepest spaces of depth d, and that $d > 2$.

We carry out a depth-reducing procedure with C7. Inspection of possibilities shows that not more than $2^i - 1$ steps are needed to find $e = e_1$ so that e_1 has (say) j deepest spaces of depth $d - 1$. (The maximum number of steps is needed in case the part of s_{d-2} in e is the only part containing s_d, and each division of s_d is contained in a separate division of s_{d-1}.) If $(d - 1) > 2$ we continue the procedure with at most $2^j - 1$ additional steps to find $e_1 = e_2$ so that e_2 is only $d - 2$ crosses deep. We see that the procedure can be continued until we find $e = e_{d-2}$ so that e_{d-2} is only $d - (d - 2) = 2$ crosses deep, and this completes the proof.

Theorem 15. Canon with respect to a variable

From any given expression, an equivalent expression can be derived so as to contain not more than two appearances of any given variable.

Proof

The proof is trivial for a variable not contained in the original expression e, and so we may confine our consideration to the case of a variable v contained in e. Now by C1 and T14

$$e = \dots \overline{\overline{vb}\; q}\; \overline{\overline{va}\; p}\; f\; \overline{vx}\; \overline{vy}\; \dots,$$

in which $a, b, \dots, p, q, \dots, x, y, \dots$, and f stand for arrangements appropriate to the expression e,

$$= \dots \overline{\overline{v}\; q}\; \overline{\overline{b}\; q}\; \overline{\overline{v}\; p}\; \overline{\overline{a}\; p}\; f\; \overline{vx}\; \overline{vy}\; \dots$$

C1,
J2,
C1 (each as often as necessary)

$$= \dots \overline{\overline{v}\; q}\; \overline{\overline{v}\; p}\; g\; \overline{vx}\; \overline{vy}\; \dots$$

calling $g = f\; \overline{\overline{a}\; p}\; \overline{\overline{b}\; q}\; \dots$

$$= \overline{\overline{\overline{p}\; \overline{q}\; \dots}\; \overline{v}}\; \overline{\overline{\overline{x}\; \overline{y}\; \dots}\; v}\; g$$

C1,
J2 (twice each)

and this completes the proof.

8

RE-UNITING THE TWO ORDERS

Content, image, and reflexion

Of any expression e, call e the content, call $\overline{e\,|}$ the image, and call $\overline{\overline{e\,|}\,|}$ the reflexion.

Since $\overline{\overline{e\,|}\,|} = e$, the act of reflexion is a return from an image to its content or from a content to its image.

Suppose e is a cross. The content of e is the content of the space in which it stands, not the content of the cross which marks the space.

In general, a content is where we have marked it, and a mark is not inside the boundary shaping its form, but inside the boundary surrounding it and shaping another form. Thus in describing a form, we find a succession,

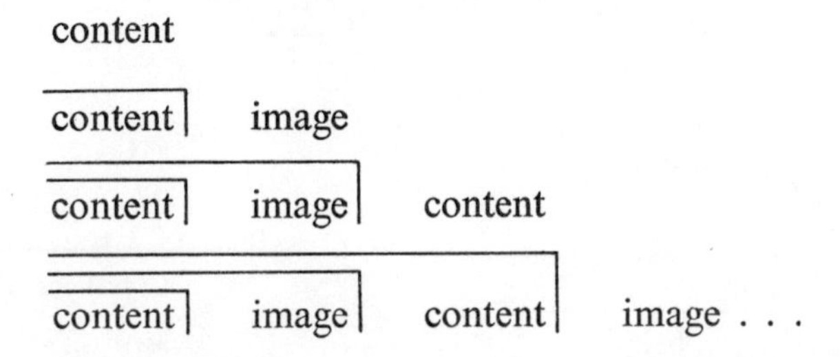

Indicative space

If s_0 is the pervasive space of e, the value of e is its value to s_0. If e is the whole expression in s_0, s_0 takes the value of e and we can call s_0 the indicative space of e.

In evaluating e we imagine ourselves in s_0 with e and thus surrounded by the unwritten cross which is the boundary to s_{-1}.

Seventh canon. Principle of relevance

If a property is common to every indication, it need not be indicated.

An unwritten cross is common to every expression in the calculus of indications and so need not be written. Similarly, a recessive value is common to every expression in the calculus of indications and also, by this principle, has no necessary indicator there.

In the form of any calculus, we find the consequences in its content and the theorems in its image.

Thus

$$\overline{\overline{\overline{}\rvert}\rvert\;\overline{}\rvert}\rvert\;\overline{}\rvert\;\overline{}\rvert \;=\; \overline{}\rvert$$

is a consequence in, and therefore in the content of, the primary arithmetic.

Demonstration

$$\begin{aligned}
&\overline{\overline{\overline{}\rvert}\rvert\;\overline{}\rvert}\rvert\;\overline{}\rvert\;\overline{}\rvert && \\
&= \overline{\overline{}\rvert}\rvert\;\overline{}\rvert\;\overline{}\rvert && \text{I2} \\
&= \overline{}\rvert\;\overline{}\rvert && \text{I2} \\
&= \overline{}\rvert && \text{I1.}
\end{aligned}$$

A consequence is acceptable because we decided the rules. All we need to show is that it follows through them.

But demonstrations of any but the simplest consequences in the content of the primary arithmetic are repetitive and tedious, and we can contract the procedure by using theorems, which are about, or in the image of, the primary arithmetic. For example, instead of demonstrating the consequence above, we can use T2.

T2 is a statement that all expressions of a certain kind, which it describes without enumeration, and of which the expression above can be recognized as an example, indicate the marked state. Its proof may be regarded as a simultaneous demonstration of all the simplifications of expressions of the kind it describes.

But the theorem itself is not a consequence. Its proof does not proceed according to the rules of the arithmetic, but follows, instead, through ideas and rules of reasoning and counting which, at this stage, we have done nothing to justify.

Thus if any person will not accept a proof, we can do no better than try another. A theorem is acceptable because what it states is evident, but we do not as a rule consider it worth recording if its evidence does not need, in some way, to be made evident. This rule is excepted in the case of an axiom, which may appear evident without further guidance. Both axioms and theorems are more or less simple statements about the ground on which we have chosen to reside.

Since the initial steps in the algebra were taken to represent theorems about the arithmetic, it depends on our point of view whether we regard an equation with variables as expressing a consequence in the algebra or a theorem about the arithmetic. Any demonstrable consequence is alternatively provable as a theorem, and this fact may be of use where the sequence of steps is difficult to find. Thus, instead of demonstrating in algebra the equation

$$\overline{\overline{\overline{a}\rceil\;\overline{b}\rceil}\rceil\;\overline{a\;\overline{c}\rceil}\rceil}\rceil = \overline{\overline{a}\rceil\;b}\rceil\;\overline{ac}\rceil \;,$$

we can prove it by arithmetic.

Call

E1 $$\overline{\overline{\overline{a}\rceil\;\overline{b}\rceil}\rceil\;\overline{a\;\overline{c}\rceil}\rceil}\rceil = x$$

and

E2 $$\overline{\overline{a}\rceil\;b}\rceil\;\overline{ac}\rceil = y.$$

Take $a = \overline{\;\;}|$. Thus

$$x = \overline{\overline{\overline{\overline{\;\;}|}|\;\overline{b\,}|}|\;\overline{\overline{\;\;}|\;\overline{c\,}|}|}| \qquad \text{substitution in E1}$$

$$= \overline{\overline{\overline{b\,}|}|\;\overline{\overline{\;\;}|\;\overline{c\,}|}|}| \qquad \text{I2}$$

$$= \overline{\overline{\overline{b\,}|}|\;\overline{\overline{\;\;}|\;\overline{b\,}|\;\overline{c\,}|}|}| \qquad \text{T2, T7}$$

$$= \overline{\overline{\;\;}|\;\overline{\overline{\;\;}|\;\overline{c\,}|}|}|\;\overline{b\,}| \qquad \text{T9}$$

$$= \overline{\overline{\;\;}|\;\overline{\overline{\;\;}|}|}|\;\overline{b\,}| \qquad \text{T2}$$

$$= \overline{b\,}| \qquad \text{I2 (twice)}$$

and

$$y = \overline{\overline{\overline{\;\;}|}|\;b\,}|\;\overline{\overline{\;\;}|\;c\,}| \qquad \text{substitution in E2}$$

$$= \overline{\overline{\overline{\;\;}|}|\;b\,}|\;\overline{\overline{\;\;}|}| \qquad \text{T2}$$

$$= \overline{b\,}| \qquad \text{I2 (twice)}$$

and so $x = y$ in this case, T7.

Now take $a =$. Thus

$$x = \overline{\overline{\overline{\;\;}|\;\overline{b\,}|}|\;\overline{\overline{c\,}|}|}| \qquad \text{substitution in E1}$$

$$= \overline{\overline{\overline{\;\;}|\;\overline{b\,}|\;\overline{c\,}|}|\;\overline{\overline{c\,}|}|}| \qquad \text{T2, T7}$$

$$= \overline{\overline{\overline{\;\;}|\;\overline{b\,}|}|\;\overline{\;\;}|}|\;\overline{c\,}| \qquad \text{T9}$$

$$= \overline{\overline{\overline{\;\;}\rvert}\rvert\;\overline{\;\;}\rvert}\rvert\;\overline{c\,}\rvert \qquad \text{T2}$$

$$= \overline{c\,}\rvert \qquad \text{I2 (twice)}$$

and

$$y = \overline{\overline{\;\;}\rvert\; b\,}\rvert\;\overline{c\,}\rvert \qquad \text{substitution in E2}$$

$$= \overline{\overline{\;\;}\rvert}\rvert\;\overline{c\,}\rvert \qquad \text{T2}$$

$$= \overline{c\,}\rvert \qquad \text{I2}$$

and so $x = y$ in this case, T7.

There is no other case, T1.

Therefore $x = y$.

By their origin, the consequences in the algebra are arithmetically valid, so we may use them as we please to shorten the proof.

Abridged proof

Take $a = \overline{\;\;}\rvert$. Thus

$$x = \overline{\overline{\overline{\overline{\;\;}\rvert}\rvert\;\overline{b\,}\rvert}\rvert\;\overline{\overline{\;\;}\rvert\;\overline{c\,}\rvert}\rvert}\rvert \qquad \text{substitution in E1}$$

$$= \overline{b\,}\rvert \qquad \text{C3, C1 (thrice)}$$

and

$$y = \overline{\overline{\overline{\;\;}\rvert}\rvert\; b\,}\rvert\;\overline{\overline{\;\;}\rvert\; c\,}\rvert \qquad \text{substitution in E2}$$

$$= \overline{b\,}\rvert \qquad \text{C3, C1 (twice)}$$

and so $x = y$ in this case, T7.

Take $a = \quad$. Thus

$$x = \overline{\overline{\overline{}|\; b}|\;\overline{\overline{c}|}|}| \qquad \text{substitution in E1}$$

$$= \overline{c}| \qquad \text{C3, C1 (twice)}$$

and

$$y = \overline{\overline{}|\; b}|\;\overline{c}| \qquad \text{substitution in E2}$$

$$= \overline{c}| \qquad \text{C3, C1}$$

and so $x = y$ in this case, T7.

There is no other case, T1.

Therefore $x = y$.

In these proofs we evidently supposed the irrelevance of variables other than the one we fixed arithmetically. It may not at first be obvious that we can ignore the possible values of the other variables, but the supposition is in fact justified in all instances (and, indeed, in all algebras), as the following proof will show.

Theorem 16. The bridge

If expressions are equivalent in every case of one variable, they are equivalent.

Let a variable v in a space s_q oscillate between the limits of its value m, n.

If the value of every other indicator in s_q is n, the oscillation of v will be transmitted through s_q and seen as a variation in the value of the boundary of s_q to s_{q-1}.

Under this condition call s_q transparent.

If the value of any other indicator in s_q is m, nothing will be transmitted through s_q.

Under this condition call s_q opaque.

The transmission from v is the alternation between transparency and opacity in s_q and in any more distant space in which this alternation can be detected. It may at any point be absorbed in transmissions from other variables in the space through which it passes. On condition that this absorption is total, call the band of space in which it occurs opaque. Under any other condition, call it transparent.

From these definitions and considerations we can see the following principle.

Eighth canon. Principle of transmission

With regard to an oscillation in the value of a variable, the space outside the variable is either transparent or opaque.

Proof of theorem 16

Let s, t be the indicative spaces of e, f respectively.

Let either of e, f contain a variable v, and let v oscillate between the limits of its value m, n.

Consider the condition under which both e and f are opaque to transmission from v. If e and f are equivalent after a change in the value of v, they were equivalent before.

Thus $e = f$ under this condition.

Consider either e or f transparent.

Suppose the oscillation of v is transmitted to one indicative space and not to the other. By selecting an appropriate value of v, we could make e not equivalent to f, and this is contrary to hypothesis. Thus if either of e or f is transparent, both are transparent.

Thus any change in the value of v is transmitted to s and t.

Therefore, if e and f are equivalent after a change in v, they were equivalent before.

Thus $e = f$ under this condition.

But, by the principle of transmission, there is no other condition.

Therefore $e = f$ under any condition, and hence in any case.

This completes the proof.

9

COMPLETENESS

We have seen that any demonstrable consequence in the algebra must indicate a provable theorem about the arithmetic. In this way consequences in the algebra may be said to represent properties of the arithmetic. In particular, they represent the properties of the arithmetic that can be expressed in forms of equation.

We can question whether the algebra is a complete or only a partial account of these properties. That is to say, we can ask whether or not every form of equation which can be proved as a theorem about the arithmetic can be demonstrated as a consequence in the algebra.

Theorem 17. Completeness

The primary algebra is complete.

That is to say, if $\alpha = \beta$ can be proved as a theorem about the primary arithmetic, then it can be demonstrated as a consequence for all α, β in the primary algebra.

We prove this theorem by induction. We first show that if all cases of $\alpha = \beta$ are algebraically demonstrable with less than a certain positive number n of distinct variables, then so is any case of $\alpha = \beta$ with n distinct variables. We then show that the condition of complete demonstrability in cases of less than n variables does in fact hold for some positive value of n.

Proof

Suppose that the demonstrability of $\alpha = \beta$ is established for all equivalent α, β containing an aggregate of less than n distinct variables.

Let a given equivalent α, β contain between them n distinct variables.

Procedure. Reduce the given α, β to their canonical forms, say α', β', with respect to a variable v.

We see in the proofs of T14 and T15 that this reduction is algebraic, so that $\alpha = \alpha'$ and $\beta = \beta'$ are both demonstrable, and that no distinct variable is added during the course of it.

By the proof of T15 we may suppose the canonical form of α to be $\overline{\overline{v}|\,A_1}|\;\overline{v\,A_2}|\;A_3$, and that of β to be $\overline{\overline{v}|\,B_1}|\;\overline{v\,B_2}|\;B_3$. Hence

E1 $$\alpha = \overline{\overline{v}|\,A_1}|\;\overline{v\,A_2}|\;A_3$$

and

E2 $$\beta = \overline{\overline{v}|\,B_1}|\;\overline{v\,B_2}|\;B_3$$

are both demonstrable. Thus

$$\overline{\overline{v}|\,A_1}|\;\overline{v\,A_2}|\;A_3 \quad = \quad \overline{\overline{v}|\,B_1}|\;\overline{v\,B_2}|\;B_3$$

is true, although we do not yet know if it is demonstrable. But by substituting constant values for v we find

E3 $$\overline{A_1}|\;A_3 \quad = \quad \overline{B_1}|\;B_3$$

E4 $$\overline{A_2}|\;A_3 \quad = \quad \overline{B_2}|\;B_3.$$

Now each of E3, E4, having at most $n - 1$ distinct variables, is demonstrable by hypothesis. Hence E1—4 are all demonstrable, and we can demonstrate

$$\alpha = \overline{\overline{v}|\,A_1}|\;\overline{v\,A_2}|\;A_3 \qquad \text{E1}$$

$$= \overline{\overline{\overline{v}|\;\overline{A_1}|}|\;\overline{v\;\overline{A_2}|}|}|\;A_3 \qquad \text{C9}$$

$$= \overline{\overline{\overline{v}|\;\overline{A_1}|\;A_3}|\;\overline{v\;\overline{A_2}|\;A_3}|}| \qquad \text{J2}$$

$$= \overline{\overline{\overline{v}\rceil \; \overline{B_1}\rceil \; B_3}\rceil \; \overline{v \; \overline{B_2}\rceil \; B_3}\rceil}\rceil \qquad \text{E3, E4}$$

$$= \beta \qquad \text{J2, C9, E2.}$$

Thus $\alpha = \beta$ is demonstrable with n variables on condition that it is demonstrable with fewer than n variables.

It remains to show that there exists a positive value of n for which $\alpha = \beta$ is demonstrable for all equivalent α, β with fewer than n variables.

It is sufficient to prove the condition for $n = 1$. Thus we need to show that if $\alpha = \beta$ contains no variable, it is demonstrable in the algebra.

If α, β contain no variable, they may be considered as expressions in the primary arithmetic.

We see in the proofs of T1–4 that all arithmetical equations are demonstrable in the arithmetic. It remains to show that they are demonstrable in the algebra.

In C3 let $a = \overline{}\rceil$ to give

$$\overline{}\rceil \; \overline{}\rceil = \overline{}\rceil$$

and this is I1.

In C1 let $a = \quad$ to give

$$\overline{\overline{}\rceil}\rceil = \quad$$

and this is I2.

Thus the initials of the arithmetic are demonstrable in the algebra, and so if $\alpha = \beta$ contains no variable it is demonstrable in the algebra.

This completes the proof.

10

INDEPENDENCE

We call the equations in a set independent if no one equation can be demonstrated from the others.

Theorem 18. Independence

The initials of the primary algebra are independent.

That is to say, given J1 as the only initial, we cannot find J2 as a consequence, and given J2 as the only initial, we cannot find J1 as a consequence.

Proof

Suppose J1 determines the only transformation allowed in the algebra. It follows from the convention of intention that no expression other than of the form $\overline{\overline{p}\rceil\ p}\rceil$ can be put into or taken out of any space.

But, in J2, r is taken out of one space and put into another, and r is not necessarily of the form $\overline{\overline{p}\rceil\ p}\rceil$.

Therefore, J2 cannot be demonstrated as a consequence of J1.

Next suppose J2 determines the only transformation allowed in the algebra.

Inspection of J2 reveals no way of eliminating any distinct variable.

But J1 eliminates a distinct variable.

Therefore, J1 cannot be demonstrated as a consequence of J2, and this completes the proof.

11

EQUATIONS OF THE SECOND DEGREE

Hitherto we have obeyed a rule (theorem 1) which requires that any given expression, in either the arithmetic or the algebra, shall be finite. Otherwise, by the canons so far called, we should have no means of finding its value.

It follows that any given expression can be reached from any other given equivalent expression in a finite number of steps. We shall find it convenient to extract this principle as a rule to characterize the process of demonstration.

Ninth canon. Rule of demonstration

A demonstration rests in a finite number of steps.

One way to see that this rule is obeyed is to count steps. We need not confine its application to any given level of consideration. In an algebraic expression each variable represents an unknown (or immaterial) number of crosses, and so it is not possible in this case to count arithmetical steps. But we can still count algebraic steps.

We may note that, according to the observation in Chapter 6 on the nature of a step, it does not matter if several counts disagree, as long as at least one count is finite.

Consider the expression

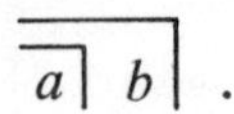

We propose now to generate a step-sequence of the following form.

$$\overline{\overline{a\,|}\;b\,|}$$

$$= \overline{\overline{a\,|}\;b\,|}\;\overline{\overline{a\,|}\;b\,|} \qquad \text{C5}$$

$$= \overline{\overline{a\,|}\;\overline{\overline{b\,|}\,|}\,|}\;\overline{\overline{a\,|}\;b\,|} \qquad \text{C1}$$

$$= \overline{\overline{\overline{\overline{a\,|}\;b\,|}\;a\,|}\;\overline{\overline{\overline{a\,|}\;b\,|}\;\overline{\overline{b\,|}\,|}\,|}\,|} \qquad \text{J2}$$

$$= \overline{\overline{\overline{\overline{a\,|}\;b\,|}\;a\,|}\;\overline{\overline{b\,|}\,|}\,|} \qquad \text{C4}$$

$$= \overline{\overline{\overline{\overline{a\,|}\;b\,|}\;a\,|}\;b\,|} \qquad \text{C1}$$

$$= \overline{\overline{\overline{\overline{a\,|}\;b\,|}\;\overline{\overline{a\,|}\;b\,|}\;a\,|}\;b\,|} \qquad \text{C5}$$

$$= \overline{\overline{\overline{\overline{a\,|}\;\overline{\overline{b\,|}\,|}\,|}\;\overline{\overline{a\,|}\;b\,|}\;a\,|}\;b\,|} \qquad \text{C1}$$

$$= \overline{\overline{\overline{\overline{\overline{\overline{a\,|}\;b\,|}\;a\,|}\;\overline{\overline{\overline{a\,|}\;b\,|}\;\overline{\overline{b\,|}\,|}\,|}\,|}\;a\,|}\;b\,|} \qquad \text{J2}$$

$$= \overline{\overline{\overline{\overline{\overline{\overline{a\,|}\;b\,|}\;a\,|}\;\overline{\overline{b\,|}\,|}\,|}\;a\,|}\;b\,|} \qquad \text{C4}$$

$$= \overline{\overline{\overline{\overline{\overline{\overline{a\,|}\;b\,|}\;a\,|}\;b\,|}\;a\,|}\;b\,|} \qquad \text{C1}$$

etc. There is no limit to the possibility of continuing the sequence, and thus no limit to the size of the echelon of alternating *a*'s and *b*'s with which $\overline{\overline{a\,|}\;b\,|}$ can be equated.

Let us imagine, if we can, that the order to begin the step-sequence is never countermanded, so that the process continues timelessly. In space this will give us an echelon without limit, of the form

$$\overline{\overline{\overline{\overline{\ldots a}|\; b}|\; a}|\; b}| \;.$$

Now, since this form, being endless, cannot be reached in a finite number of steps from $\overline{\overline{a}|\; b}|$, we do not expect it to express, necessarily, the same value as $\overline{\overline{a}|\; b}|$. But we can, by means of an exhaustive examination of possibilities, ascertain what values it might take in the various cases of *a*, *b*, and compare them with those of the finite expression.

Re-entry

The key is to see that the crossed part of the expression at every even depth is identical with the whole expression, which can thus be regarded as re-entering its own inner space at any even depth. Thus

$$f = \overline{\overline{\overline{\overline{\ldots a}|\; b}|\; a}|\; b}|$$

E1 $$= \overline{\overline{fa}|\; b}| \;.$$

We can now find, by the rule of dominance, the values which *f* may take in each possible case of *a*, *b*.

$$\overline{\overline{fa}|\; b}| = f \qquad \text{E1}$$

$$\overline{\overline{fm}|\; m}| = n$$

$$\overline{\overline{fm}|\; n}| = m$$

$$\overline{\overline{fn}|\; m}| = n$$

$$\overline{\overline{fn}|\; n}| = m \text{ or } n.$$

For the last case suppose $f = m$. Then

$$\overline{\overline{mn}\rceil \; n}\rceil \; = \; m$$

and so E1 is satisfied. Now suppose $f = n$. Then

$$\overline{\overline{nn}\rceil \; n}\rceil \; = \; n$$

and so E1 is again satisfied. Thus the equation, in this case, has two solutions.

It is evident, then, that, by an unlimited number of steps from a given expression e, we can reach an expression e' which is not equivalent to e.

We see, in such a case, that the theorems of representation no longer hold, since the arithmetical value of e' is not, in every possible case of a, b, uniquely determined.

Indeterminacy

We have thus introduced into e' a degree of indeterminacy in respect of its value which is not (as it was in the case of indeterminacy introduced merely by cause of using independent variables) necessarily resolved by fixing the value of each independent variable. But this does not preclude our equating such an expression with another, provided that the degree of indeterminacy shown by each expression is the same.

Degree

We may take the evident degree of this indeterminacy to classify the equation in which such expressions are equated. Equations of expressions with no re-entry, and thus with no unresolvable indeterminacy, will be called equations of the first degree, those of expressions with one re-entry will be called of the second degree, and so on.

It is evident that J1 and J2 hold for all equations, whatever their degree. It is thus possible to use the ordinary procedure of demonstration (outlined in Chapter 6) to verify an equation of degree > 1. But we are denied the procedure (outlined in Chapter 8) of referring to the arithmetic to confirm a demonstration of any such equation, since the excursion to infinity undertaken to produce it has denied us our former access to a complete knowledge of where we are in the form. Hence it was necessary to extract, before departing, the rule of demonstration, for this now becomes, with the rule of dominance, a guiding principle by which we can still find our way.

Imaginary state

Our loss of connexion with the arithmetic is illustrated by the following example. Let

E2 $$f_2 = \overline{\overline{f_2|}\,|} \,,$$

E3 $$f_3 = \overline{f_3|} \,.$$

Plainly, each of E2, E3 can be represented, in arithmetic, by equating either f with the same infinite expression, thus

$$f_2, f_3 = \overline{\overline{\overline{\;\ldots\;|}\,|}\,|} \,.$$

But equally plainly, whereas E2 is open to the arithmetical solutions $\overline{\ \ }|$ or $\quad$, each of which satisfies it without contradiction, E3 is satisfied by neither of these solutions, and cannot, thereby, express the same value as E2. And since $\overline{\ \ }|$ and $\quad$ represent the only states of the form hitherto envisaged, if we wish to pretend that E3 has a solution, we must allow it to have a solution representing an imaginary state, not hitherto envisaged, of the form.

Time

Since we do not wish, if we can avoid it, to leave the form, the state we envisage is not in space but in time. (It being possible to enter a state of time without leaving the state of space in which one is already lodged.)

One way of imagining this is to suppose that the transmission of a change of value through the space in which it is represented takes time to cover distance. Consider a cross

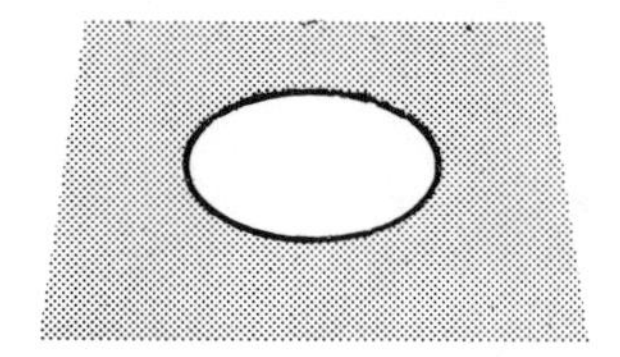

in a plane. An indication of the marked state is shown by the shading.

Now suppose the distinction drawn by the cross to be destroyed by a tunnel under the surface in which it appears. In Figure 1 we see the results of such destruction at intervals t_1, t_2, . . .

Frequency

If we consider the speed at which the representation of value travels through the space of the expression to be constant, then the frequency of its oscillation is determined by the length of the tunnel. Alternatively, if we consider this length to be constant, then the frequency of the oscillation is determined by the speed of its transmission through space.

Velocity

We see that once we give the transmission of an indication of value a speed, we must also give it a direction, so that it becomes a velocity. For if we did not, there would be nothing to stop the propagation proceeding as represented to t_4 (say) and then continuing towards the representation shown in t_3 instead of that shown in t_5.

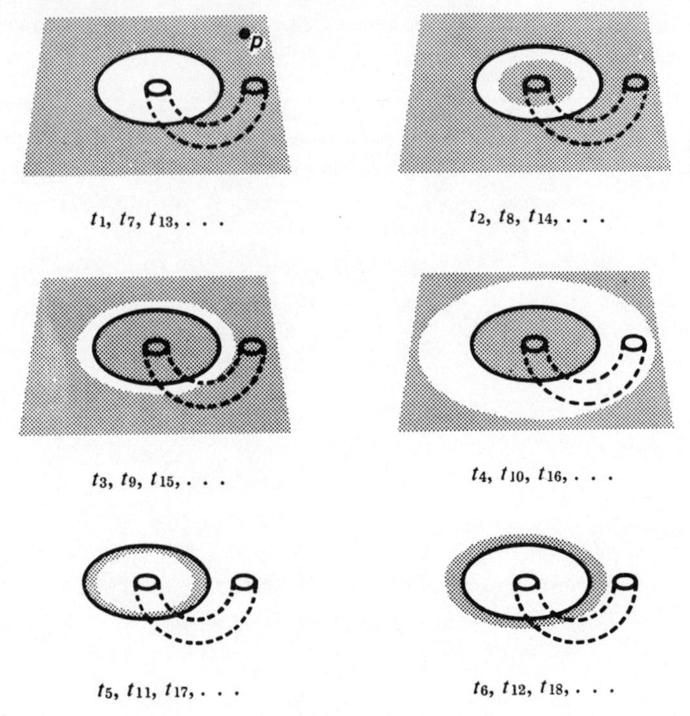

Figure 1

Function

We shall call an expression containing a variable v alternatively a function of v. We thus see expressions of value or functions of variables, according to from which point of view we regard them.

Oscillator function

In considering the indications of value at the point p in Figure 1, we have, in time, a succession of square waves of a given frequency.

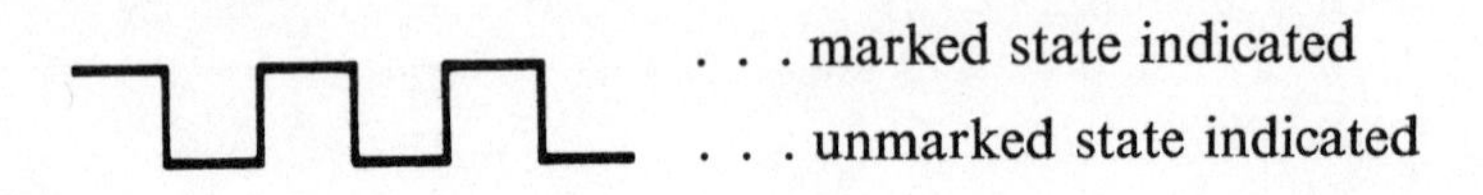

Suppose we now arrange for all the relevant properties of the point p in Figure 1 to appear in two successive spaces of expression, thus.

$$\overline{p\,|}\;p$$

We could do this by arranging similarly undermined distinctions in each space, supposing the speed of transmission to be constant throughout. In this case the superimposition of the two square waves in the outer space, one of them inverted by the cross, would add up to a continuous representation of the marked state there.

Real and imaginary value

The value represented at (or by) the point (or variable) p, being indeterminate in space, may be called imaginary in relation with the form. Nevertheless, as we see above, it is real in relation with time and can, in relation with itself, become determinate in space, and thus real in the form.

We have considered thus far a graphical representation of E3. We will now consider E1 and its limiting case E2 on similar lines.

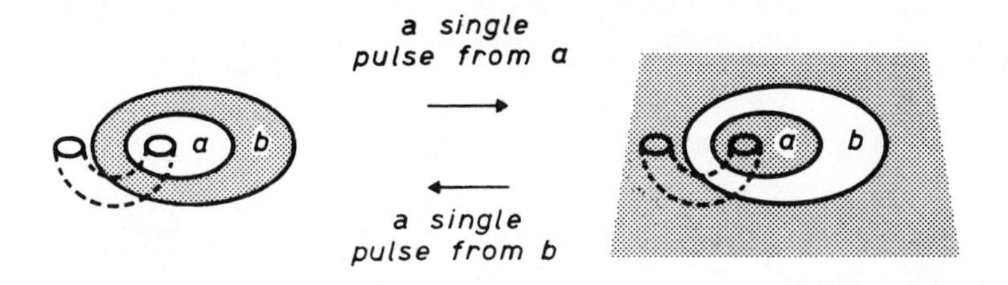

Figure 2

Memory function

The present value of the function f in E1 may depend on its past value, and thus on past values of a and b. In effect, when a, b both indicate the unmarked state, it remembers which of them last indicated the marked state. If a, then $f = m$. If b, then $f = n$.

Subversion

A way to make the set-up illustrated in Figure 2 behave exactly like the f in E1, is to arrange that effective transmission through the tunnel shall be only from outside to inside. We shall call such a partial destruction of the distinctive properties of constants a subversion.

We may note that, if we wish to avail ourselves of the memory property of f, where f is an evenly subverted function, certain transformations, allowable in the case of an expression without this property, must be avoided. We may, for example, allow

$$\overline{\overline{\overline{\overline{a}\ \overline{b}}\ f}\ c} \rightarrow \overline{\overline{fa}\ \overline{fb}\ c} \qquad \textbf{J2, C1}$$

but must avoid

$$\overline{\overline{\overline{\overline{a}\ \overline{b}}\ f}\ c} \rightarrow \overline{\overline{\overline{\overline{a}\ \overline{b}}\ fc}\ c} \qquad \textbf{C2}$$

since the latter transformation is from an expression by which an indication of the marked state by c can be reliably remembered to an expression in which the memory is apparently lost.

Time in finite expressions

The introduction of time into our deliberations did not come as an arbitrary choice, but as a necessary measure to further the inquiry.

The degree of necessity of a measure adopted is the extent of its application. The measure of time, as we have introduced it here, can be seen to cover, without inconsistency, all the representative forms hitherto considered.

This can be illustrated by reconsidering E1. Here we can test the use of the concept of time by finding whether it leads to the same answer (i.e. whether it leads to the same memory of dominant states of a, b) in the expanded version of f as it does in the contracted version in Figure 2. For the purpose of illustration, we shall consider a finite expression first.

It is seen from Figure 3 that such a finite expression is stable in one condition, and has a finite memory of the other, of

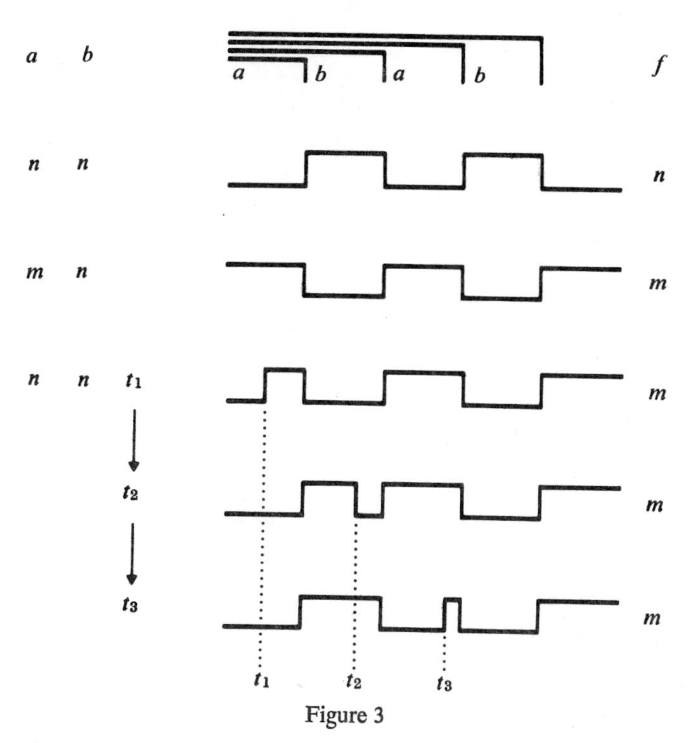

Figure 3

duration proportional to the degree of its extension. It is plain that an endless extension of the echelon allows an endless memory of either condition, so that the concept of time is a key by which the contracted and expanded forms of f in E1 are made patent to one another.

A condition of special interest emerges if the dominant pulse from a is of sufficiently short duration. In this condition the expression emits a wave train of finite length and duration, as illustrated in Figure 4.

The duration of the wave train, the frequency of its components, etc, depend on the nature and extent of the expression

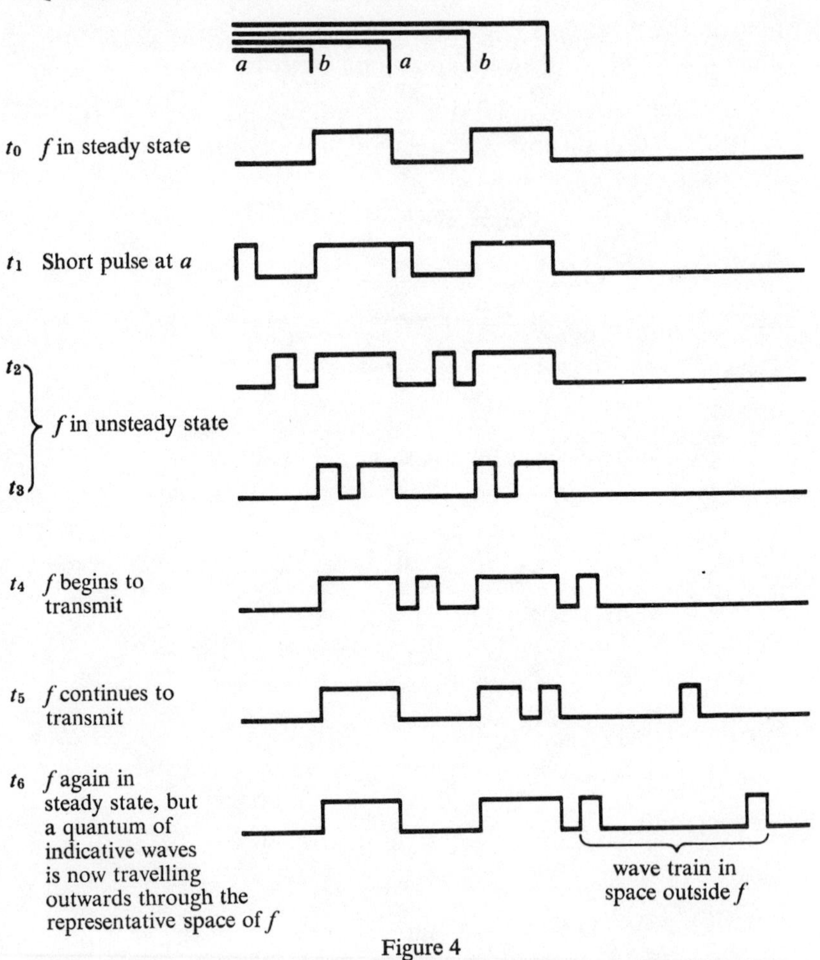

Figure 4

from which it is emitted. From an infinitely extended expression comes a potentially endless emission, and here again, the two ways (contracted or expanded) of expressing E1 in relation with time give the same answer. Without the key of time, only the contracted expression makes sense.

Crosses and markers

Consider the case where the expression in E1 represents a part of a larger expression. It now becomes necessary not only to indicate where a re-insertion takes place, but also to designate the part of the expression re-inserted. Since the whole is no longer the part re-inserted, it will be necessary in each case either to name the part re-inserted or to indicate it by direct connexion.

The latter is less cumbersome. Thus we can rewrite the expression in E1

$$\overline{\underline{\left\lfloor\, \overline{a\,}\!\rceil\; b \,\right\rfloor}}\!\Big|$$

so that it can be placed, without ambiguity, within a larger expression.

In a simple subverted expression of this kind neither of the non-literal parts are, strictly speaking, crosses, since they represent, in a sense, the same boundary. It is convenient, nevertheless, to refer to them separately, and for this purpose we call each separate non-literal part of any expression a marker. Thus a cross is a marker, but a marker need not be a cross.

Modulator function

We have seen that functions of the second degree can either oscillate or remember. If we are prepared to write an equation of degree >2 we can find a function which will not only remember, but count.

A way of picturing counting is to consider it as the contrary of remembering. A memory function remembers the same response to the same signal: a counting function counts it different each time.

Another way to picture counting is as a modulation of a wave structure. This is the way we shall picture it here.

The simplest modulation is to a wave structure of half the frequency of the original. To achieve this with a function using only real values, we need eight markers, thus.

E4 $f =$ 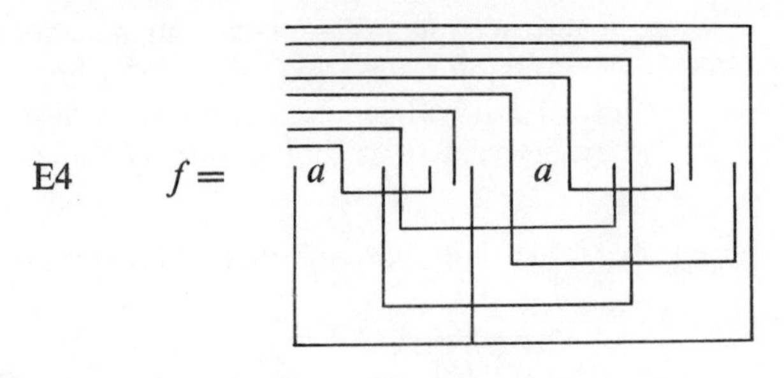

If the wave structure of a is then that of f will be or , depending on how the expression is originally set before a starts to oscillate.

We are now in difficulties through attempting to write in two dimensions what is clearly represented in three. We ought to be writing in three dimensions. We can at least devise a better system of drawing three-dimensional representations in two.

Let a marker be represented by a vertical stroke, thus.

|

Let what is under the marker be seen to be so by lines of connexion, called leads, thus.

Let the value indicated by the marker be led from the marker by a lead, which may, in the expression, divide to be entered under other markers. Now, for example, the expression

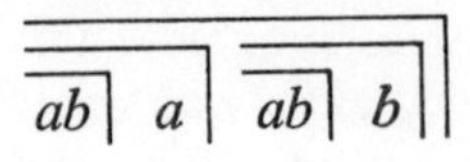

can be represented thus.

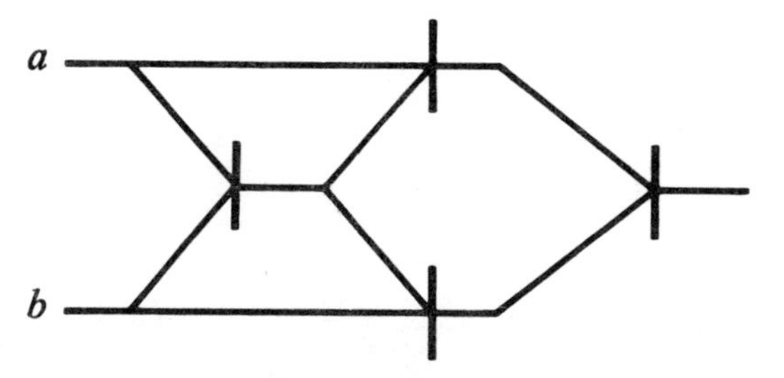

Transfigured in this way, E4 appears in a form

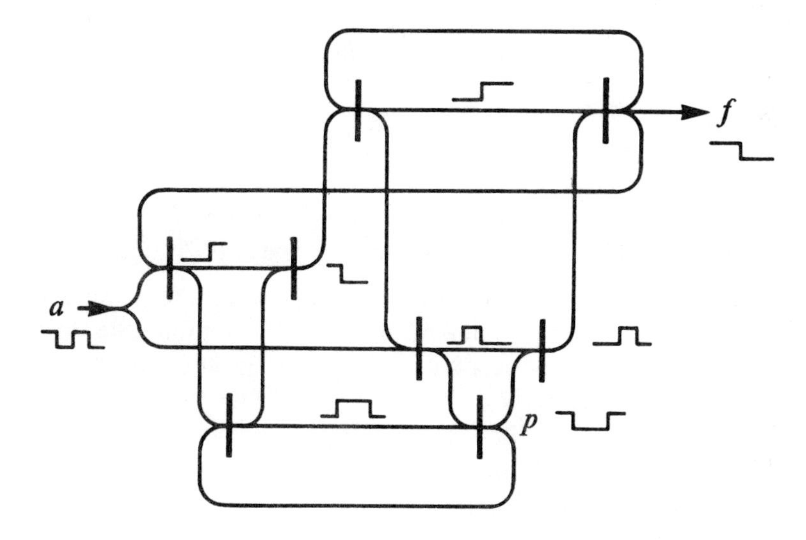

in which it is easier to follow how the wave structure of *a* is taken apart and recombined to give that of *f*.

We see that the wave structure at *p* constitutes a similar modulation with the phase displaced. By using imaginary components of some wave structures, it is possible to obtain the wave structure at *p* with only six markers. This is illustrated in the following equation.

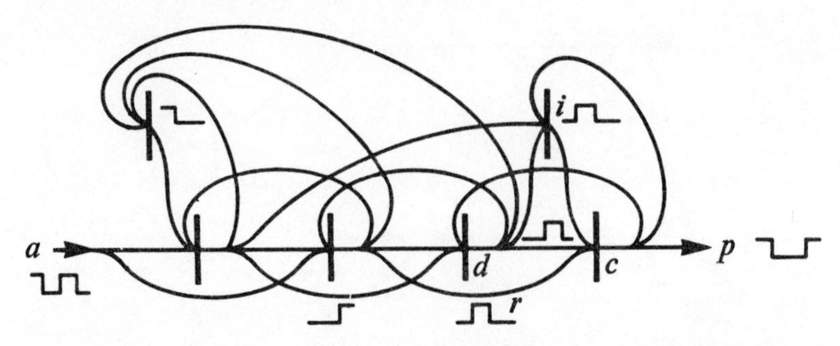

Here, although the real wave structure at *i* is identical with that at *r*, the imaginary component at *i* ensures that the memory in markers *c* and *d* is properly set. Similar considerations apply to other memories in the expression.

Coda

At this point, before we have gone so far as to forget it, we may return to consider what it is we are deliberating.

We are, and have been all along, deliberating the form of a single construction (commanded on p 3), notably the first distinction. The whole account of our deliberations is an account of how it may appear, in the light of various states of mind which we put upon ourselves.

By the canon of expanding reference (p 10), we see that the account may be continued endlessly.

This book is not endless, so we have to break it off somewhere. We now do so here with the words

and so on.

Before departing, we return for a last look at the agreement with which the account was opened.

12

RE-ENTRY INTO THE FORM

The conception of the form lies in the desire to distinguish.

Granted this desire, we cannot escape the form, although we can see it any way we please.

The calculus of indications is a way of regarding the form.

We can see the calculus by the form and the form in the calculus unaided and unhindered by the intervention of laws, initials, theorems, or consequences.

The experiments below illustrate one of the indefinite number of possible ways of doing this.

We may note that in these experiments the sign

$$=$$

may stand for the words

is confused with.

We may also note that the sides of each distinction experimentally drawn have two kinds of reference.

The first, or explicit, reference is to the value of a side, according to how it is marked.

The second, or implicit, reference is to an outside observer. That is to say, the outside is the side from which a distinction is supposed to be seen.

First experiment

In a plane space, draw a circle.

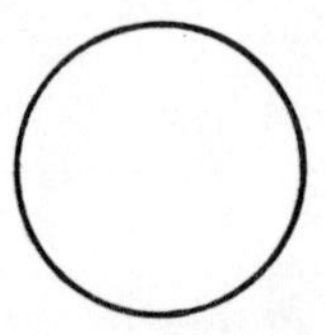

Let a mark m indicate the outside of the circumference.

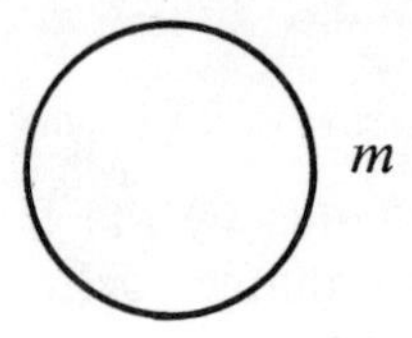

Let no mark indicate the inside of the circumference.

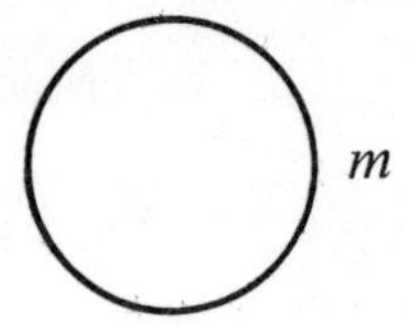

Let the mark m be a circle.

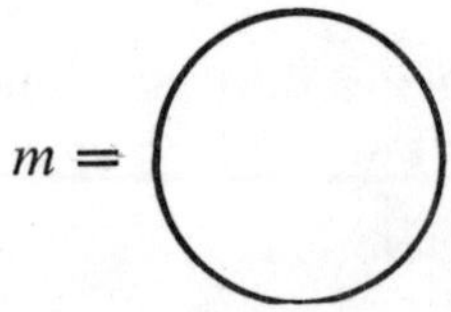

$m =$

Re-enter the mark into the form of the circle.

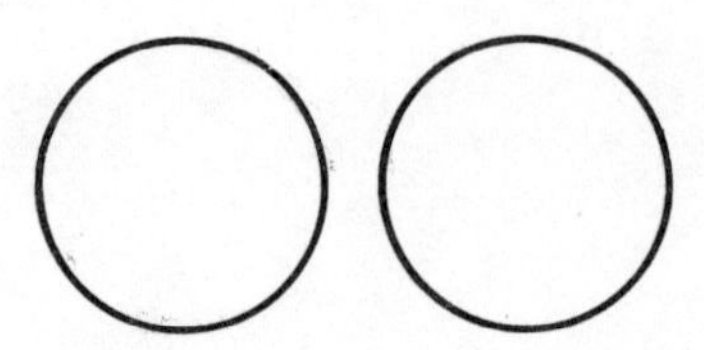

Now the circle and the mark cannot (in respect of their relevant properties) be distinguished, and so

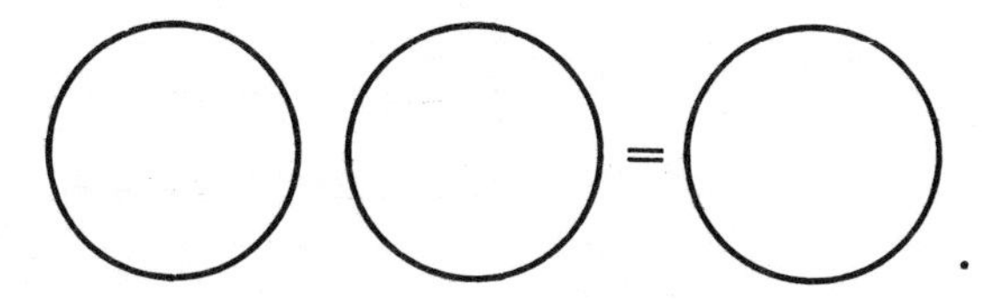

Second experiment

In a plane space, draw a circle.

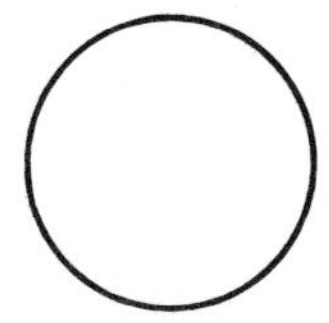

Let a mark m indicate the inside of the circumference.

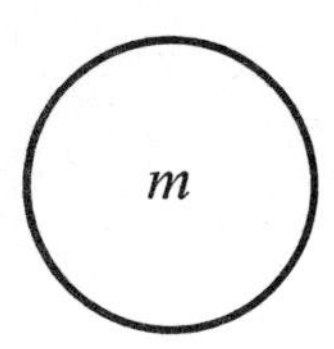

Let no mark indicate the outside of the circumference.

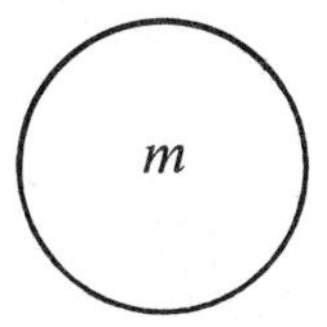

Let the value of a mark be its value to the space in which it stands. That is to say, let the value of a mark be to the space outside the mark.

Now the space outside the circumference is unmarked.

Therefore, by valuation,

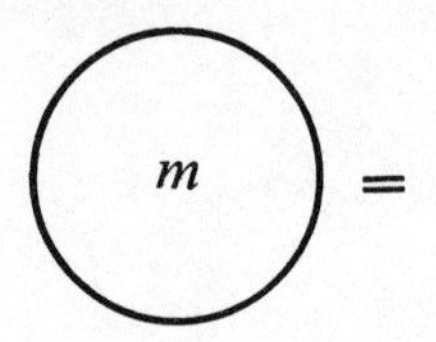

Let the mark m be a circle.

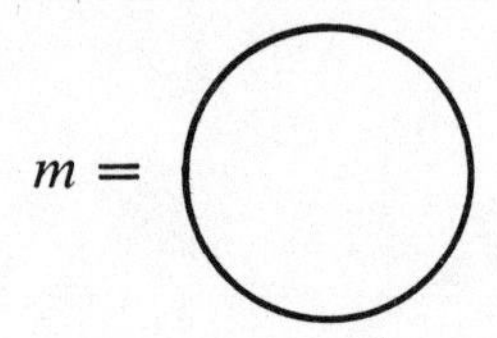

Re-enter the mark into the form of the circle.

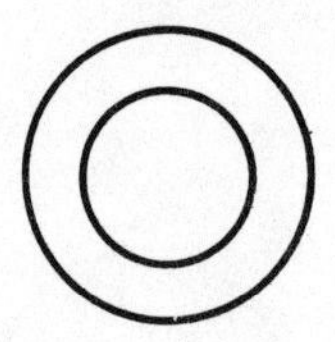

Now, by valuation,

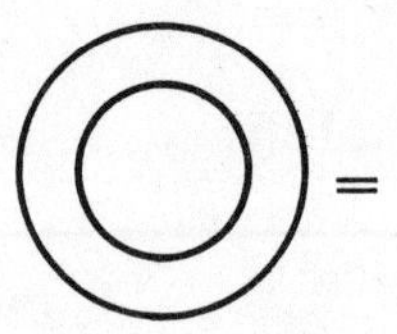

Third experiment

In a plane space, draw a circle.

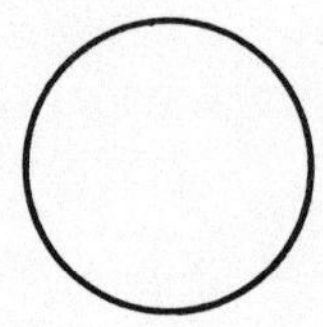

Let a mark *m* indicate the outside of the circumference.

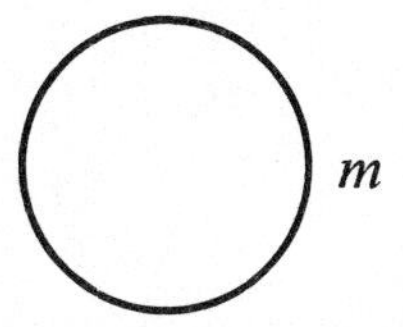

Let a similar mark *m* indicate the inside of the circumference.

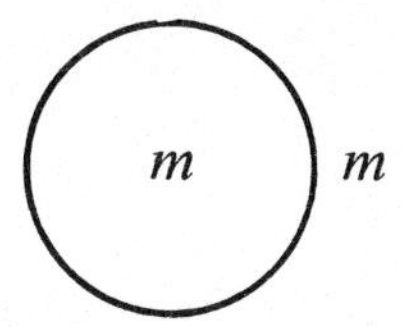

Now, since a mark *m* indicates both sides of the circumference, they cannot, in respect of value, be distinguished.

Again let the mark *m* be a circle.

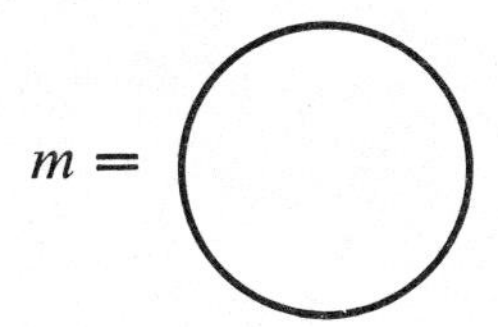

Re-enter the mark into the form of the circle.

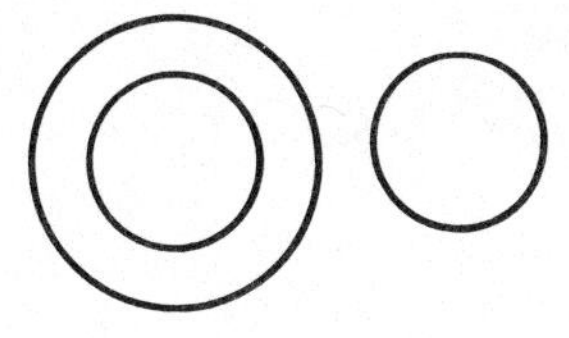

Now, because of identical markings, the original circle cannot distinguish different values.

Therefore, it is not, in this respect, a distinction.

Therefore it may be deleted without loss or gain to the space in which it stands.

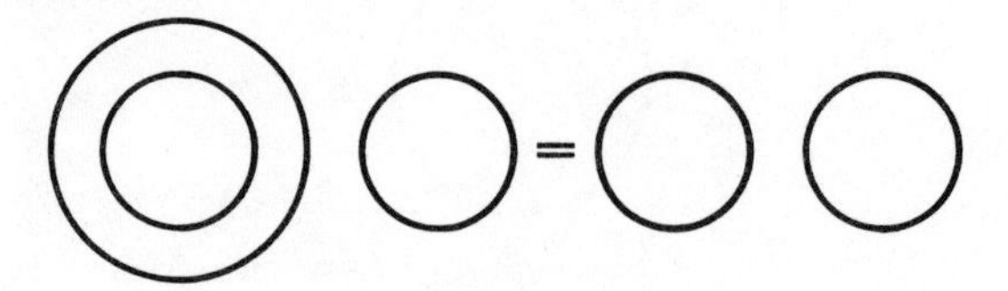

But we found in the first experiment that

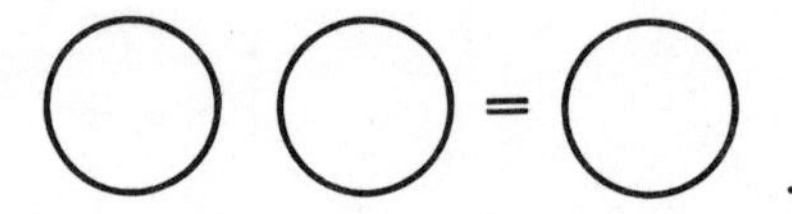
.

Therefore,

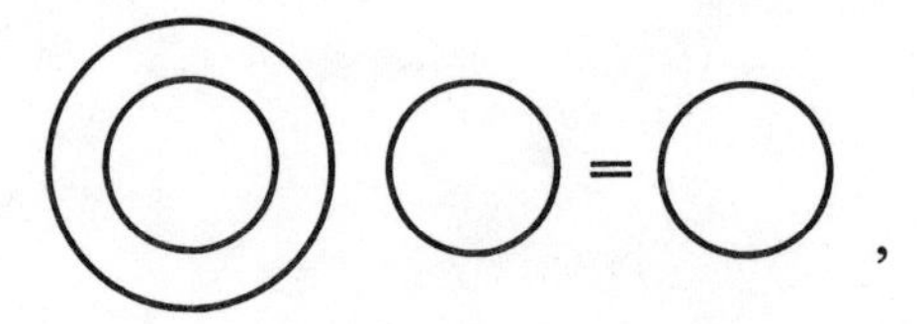
,

and this is not inconsistent with the finding of the second experiment that

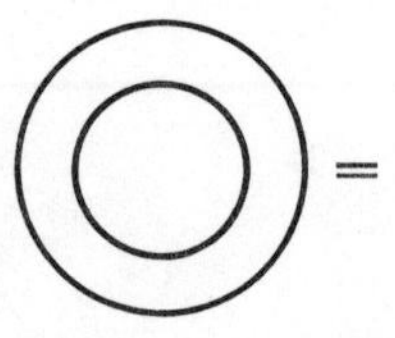
,

since we have done here in two steps which was done there in one.

Fourth experiment

In a plane space, draw a circle.

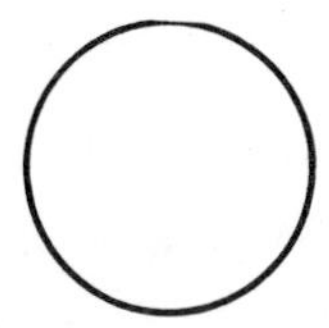

Let the outside of the circumference be unmarked.

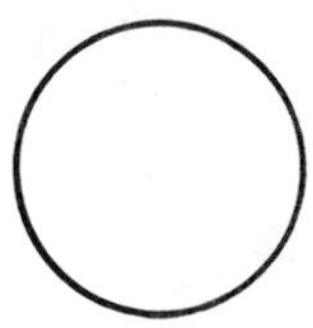

Let the inside of the circumference be unmarked.

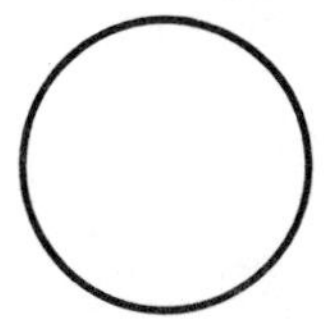

But we saw in the first experiment that

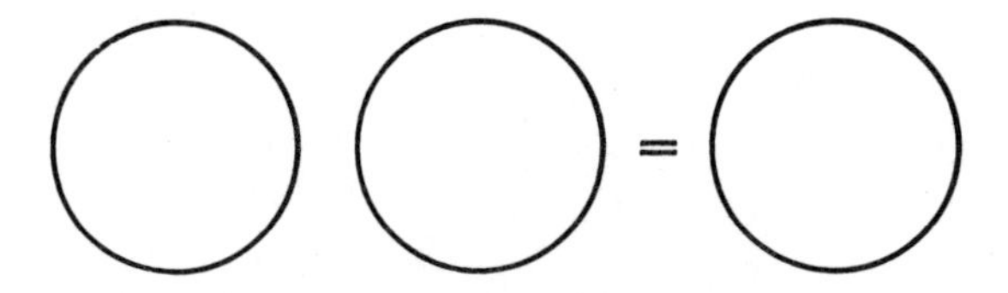

and that therefore, by reversing the purifying procedure there,

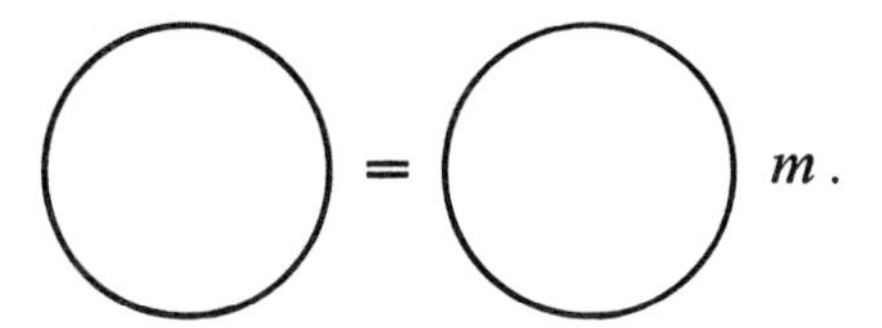

The value of a circumference to the space outside must be, therefore, the value of the mark, since the mark now distinguishes this space.

An observer, since he distinguishes the space he occupies, is also a mark.

In the experiments above, imagine the circles to be forms and their circumferences to be the distinctions shaping the spaces of these forms.

In this conception a distinction drawn in any space is a mark distinguishing the space. Equally and conversely, any mark in a space draws a distinction.

We see now that the first distinction, the mark, and the observer are not only interchangeable, but, in the form, identical.

NOTES

Chapter 1

Although it says somewhat more, all that the reader needs to take with him from Chapter 1 are the definition of distinction as a form of closure, and the two axioms which rest with this definition.

Chapter 2

It may be helpful at this stage to realize that the primary form of mathematical communication is not description, but injunction. In this respect it is comparable with practical art forms like cookery, in which the taste of a cake, although literally indescribable, can be conveyed to a reader in the form of a set of injunctions called a recipe. Music is a similar art form, the composer does not even attempt to describe the set of sounds he has in mind, much less the set of feelings occasioned through them, but writes down a set of commands which, if they are obeyed by the reader, can result in a reproduction, to the reader, of the composer's original experience.

Where Wittgenstein says [4, proposition 7]

> whereof one cannot speak,
> thereof one must be silent

he seems to be considering descriptive speech only. He notes elsewhere that the mathematician, descriptively speaking, says nothing. The same may be said of the composer, who, if he were to attempt a *description* (i.e. a limitation) of the set of ecstasies apparent *through* (i.e. unlimited by) his *composition*, would fail miserably and necessarily. But neither the composer nor the mathematician must, for this reason, be silent.

In his introduction to the *Tractatus*, Russell expresses what thus seems to be a justifiable doubt in respect of the rightness of Wittgenstein's last proposition when he says [p 22]

> what causes hesitation is the fact that, after all, Mr Wittgenstein manages to say a good deal about what cannot be said, thus suggesting to the sceptical reader that possibly there may be some loophole through a hierarchy of languages, or by some other exit.

The exit, as we have seen it here, is evident in the injunctive faculty of language.

Even natural science appears to be more dependent upon injunction than we are usually prepared to admit. The professional initiation of the man of science consists not so much in reading the proper textbooks, as in obeying injunctions such as 'look down that microscope'. But it is not out of order for men of science, having looked down the microscope, now to describe to each other, and to discuss amongst themselves, what they have seen, and to write papers and textbooks describing it. Similarly, it is not out of order for mathematicians, each having obeyed a given set of injunctions, to describe to each other, and to discuss amongst themselves, what they have seen, and to write papers and textbooks describing it. But in each case, the description is dependent upon, and secondary to, the set of injunctions having been obeyed first.

When we attempt to realize a piece of music composed by another person, we do so by *illustrating*, to ourselves, with a musical instrument of some kind, the composer's commands. Similarly, if we are to realize a piece of mathematics, we must find a way of illustrating, to ourselves, the commands of the mathematician. The normal way to do this is with some kind of scorer and a flat scorable surface, for example a finger and a tide-flattened stretch of sand, or a pencil and a piece of paper. Taking such an aid to illustration, we may now begin to carry out the commands in Chapter 2.

First we may illustrate a form, such as a circle or near-circle. A flat piece of paper, being itself illustrative of a plane surface, is a useful mathematical instrument for this purpose, since we happen to know that a circle in such a space does in fact draw a distinction. (If, for example, we had chosen to write upon the surface of a torus, the circle might not have drawn a distinction.)

When we come to the injunction

let there be a form distinct from the form

we can illustrate it by taking a fresh piece of paper (or another stretch of sand). Now, in this separate form, we may illustrate the command

let the mark of distinction be copied
out of the form into such another form.

It is not necessary for the reader to confine his illustrations to the commands in the text. He may wander at will, inventing his own illustrations, either consistent or inconsistent with the textual commands. Only thus, by his own explorations, will he come to see distinctly the bounds or laws of the world from which the mathematician is speaking. Similarly, if the reader does not follow the argument at any point, it is never necessary for him to remain stuck at that point until he sees how to proceed. We cannot fully understand the beginning of anything until we see the end. What the mathematician aims to do is to give a complete picture, the order *of what* he presents being essential, the order *in which* he presents it being to some degree arbitrary. The reader may quite legitimately change the arbitrary order as he pleases.

We may distinguish, in the essential order, *commands*, which call something into being, conjure up some order of being, call to order, and which are usually carried in permissive forms such as

let there be so-and-so,

or occasionally in more specifically active forms like

drop a perpendicular;

names, given to be used as reference points or tokens; in relation with the operation of *instructions*, which are designed to take effect within whatever universe has already been commanded or called to order. The institution or ceremony of naming is usually carried in the form

call so-and-so such-and-such,

and the call may be transmitted in both directions, as with the sign =, so that by calling so-and-so such-and-such we may also call such-and-such so-and-so. Naming may thus be considered to be without direction, or, alternatively, pan-directional. By contrast, instruction is directional, in that it demands a crossing from a state or condition, with its own name, to a different state or condition, with another name, such that the name of the former may not be called as a name of the latter.

The more important structures of command are sometimes called canons. They are the ways in which the guiding injunctions appear to group themselves in constellations, and are thus by no means independent of each other. A canon bears the distinction of being outside (i.e. describing) the system under construction, but a command to construct (e.g. 'draw a distinction'), even though it may be of central importance, is not a canon. A canon is an order, or set of orders, to permit or allow, but not to construct or create.

The instructions which are to take effect, within the creation and its permission, must be distinguished as those in the actual text of calculation, designated by the constants or *operators* of the calculus, and those in the context, which may themselves be instructions to name something with a particular name so that it can be referred to again without redescription.

Later on (Chapter 4) we shall come to consider what we call the proofs or justifications of certain statements. What we shall be showing, here, is that such statements are implicit in, or follow from, or are permitted by, the canons or standing orders hitherto convened or called to presence. Thus, in the structure of a proof, we shall find injunctions of the form consider such-and-such,

suppose so-and-so,

which are not commands, but *invitations* or *directions* to a *way* in which the implication can be clearly and wholly followed.

In conceiving the calculus of indications, we begin at a point of such *degeneracy* as to find that the ideas of description, indication, name, and instruction can amount to the same thing. It is of some importance for the reader to realize this for himself, or he will find it difficult to understand (although he may follow) the argument (p 5) leading to the second primitive equation.

In the command

let the crossing be to the
state indicated by the token

we at once make the token doubly meaningful, first as an instruction to cross, secondly as an indicator (and thus a name) of where the crossing has taken us. It was an open question, before obeying this command, whether the token would carry an indication at all. But the command determines without ambiguity the state to which the crossing is made and thus, without ambiguity, the indication which the token will henceforth carry.

This double carry of name-with-instruction and instruction-with-name is usually referred to (in the language of mathematics) as a structure in which ideas or meanings *degenerate*. We may also refer to it (in the language of psychology) as a place where the ideas *condense* in one symbol. It is this condensation which gives the symbol its *power*. For in mathematics, as in other disciplines, the power of a system resides in its elegance (literally, its capacity to pick out or elect), which is achieved by condensing as much as is needed into as little as is needed, and so making that little as free from irrelevance (or from elaboration) as is allowed by the necessity of writing it out and reading it in with ease and without error.

We may now helpfully distinguish between an elegance in

the calculus, which can make it easy to use, and an elegance in the descriptive context, which can make it hard to follow. We are accustomed, in ordinary life, to having indications of what to do confirmed in several different ways, and when presented with an injunction, however clear and unambiguous, which, stripped to its bare minimum, indicates what to do once and in one way only, we might refuse it. (We may consider how far, in ordinary life, we must observe the spirit rather than the letter of an injunction, and must develop the habitual capacity to interpret any injunction we receive by screening it against other indications of what we ought to do. In mathematics we have to unlearn this habit in favour of accepting an injunction literally and at once. This is why an author of mathematics must take such great pains to make his injunctions mutually permissive. Otherwise these pains, which rightly rest with the author, will fall with sickening import upon the reader, who, by virtue of his relationship with respect to the author, may be in no position to accept them.)

The second of the two primitive equations of the primary arithmetic can be derived less elegantly, but in a way that is possibly easier to follow, by allowing substitution prematurely.

Suppose we indicate the marked state by a token *m*, and, as before, let the absence of a token indicate the unmarked state.

Let a bracket round any indicator indicate, in the space outside the bracket, the state other than that indicated inside the bracket.

Thus

$$\enclose{circle}{m} =$$

and

$$\enclose{circle}{\quad} = m.$$

Substituting, we find

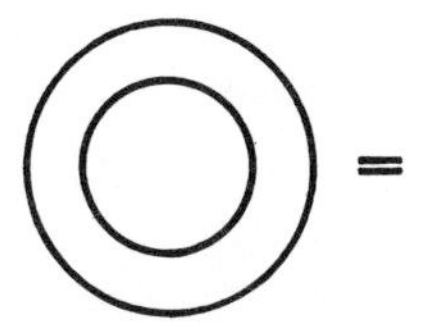

which is the second primitive equation.

The condition that one of the primary states shall be nameless is mandatory for this elimination.

The first primitive equation can also be derived a different way.

Imagine a blind animal able only to distinguish inside from outside. A space with what appears to us as a number of distinct insides and one outside, such as

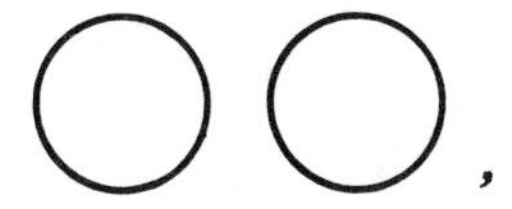

will appear to it, upon exploration, to be indistinguishable from

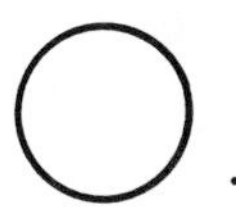

The ideas described in the text at this point do not go beyond what this animal can find out for itself, and so in its world, such as it is,

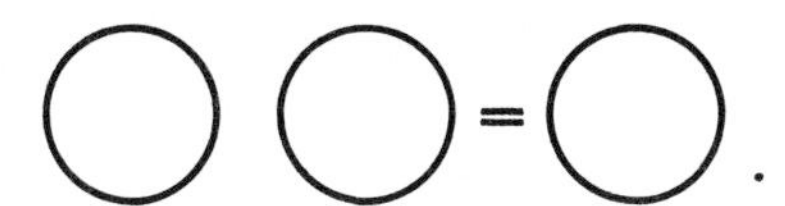

We may note that even if this animal can count its crossings, it still will not be able to distinguish two divisions from one, although it will now have an alternative way of distinguishing inside from outside which no longer depends on knowing which is which.

Reconsidering the first command,

draw a distinction,

we note that it may equally well be expressed in such ways as

let there be a distinction,

find a distinction,

see a distinction,

describe a distinction,

define a distinction,

or

let a distinction be drawn,

for we have here reached a place so primitive that active and passive, as well as a number of other more peripheral opposites, have long since condensed together, and almost any form of words will suggest more categories than there really are.

Chapter 3

The hypothesis of simplification is the first *overt* convention that is put to use before it has been justified. But it has a precursor in the injunction 'let a state indicated by an expression be the value of the expression' in the last chapter, which allows value to an expression only in case not less and not more than one state is indicated by the expression. The use of both the injunction and the convention are eventually justified in the theorems of representation. Other cases of delayed justification will be found later, a notable example being theorem 16.

We may ask why we do not justify such a convention at once when it is given. The answer, in most cases, is that the justification (although valid) would be meaningless until we had first become acquainted with the *use* of the principle which requires justifying. In other words, before we can reasonably justify a deep lying principle, we first need to be familiar with how it works.

We might suppose this practice of deferred justification to be operative elsewhere. It is a notable fact that in mathematics very few *useful* theorems remain unproved. By 'useful' I do not necessarily mean with practical application outside mathematics. A theorem can be useful mathematically, for example to justify another theorem.

One of the most 'useless' theorems in mathematics is Goldbach's conjecture. We do not frequently find ourselves saying 'if only we knew that every even number greater than 2 could be represented as a sum of two prime numbers, we should be able to show that . . .' D J Spencer Brown, in a private communication, suggested that their apparent uselessness is not exactly a reason why such theorems cannot be proved, but is a reason for supposing that if a valid proof were given today, nobody would recognize it as such, since nobody is yet *familiar* with the *ground* on which such a proof would rest. I shall have more to say about this in the notes to Chapters 8 and 11.

Chapter 4

In all mathematics it becomes apparent, at some stage, that we have for some time been following a rule without being consciously aware of the fact. This might be described as the use of a *covert* convention. A recognizable aspect of the advancement of mathematics consists in the advancement of the consciousness of what we are doing, whereby the covert becomes overt. Mathematics is in this respect psychedelic.

The nearer we are to the beginning of what we set out to achieve, the more likely we are to find, there, procedures which have been adopted without comment. Their use can be considered as the presence of an arrangement in the absence of an agreement. For example, in the statement and proof of theorem 1 it is arranged (although not agreed) that we shall write on a plane surface. If we write on the surface of a torus the theorem is not true. (Or to make it true, we must be more explicit.)

Notes

The fact that men have for centuries used a plane surface for writing means that, at this point in the text, both author and reader are ready to be conned into the assumption of a plane writing surface without question. But, like any other assumption, it is not unquestionable, and the fact that we can question it here means that we can question it elsewhere. In fact we have found a common but hitherto unspoken assumption underlying what is written in mathematics, notably a plane surface (more generally, a surface of genus 0, although we shall see later (pp 102 sq) that this further generalization forces us to recognize another hitherto silent assumption). Moreover, it is now evident that if a different surface is used, what is written on it, although identical in marking, may be not identical in meaning.

In general there is an order of precedence amongst theorems, so that theorems which can be proved more easily with the help of other theorems are placed so as to be proved after such other theorems. This order is not rigid. For example, having proved theorem 3, we use what we found in the proof to prove theorem 4. But theorems 3 and 4 are symmetrical, their order depending only on whether we wish to proceed from simplicity to complexity or from complexity to simplicity. The reader might try, if he wishes, to prove theorem 4 first without the aid of theorem 3, after which he will be able to prove theorem 3 analogously to the way theorem 4 is proved in the text.

It will be observed that the symbolic representation of theorem 8 is less strong than the theorem itself. The theorem is consistent with

$$\overline{\overline{p\,}|\;\; pq\,}| \;=\; \qquad ,$$

whereas we prove the weaker version

$$\overline{\overline{p\,}|\;\; p\,}| \;=\; \qquad .$$

The stronger version is plainly true, but we shall find that we are able to demonstrate it as a consequence in the algebra. We therefore prove, and use as the first algebraic initial, the weaker version.

In theorem 9 we see the difference between our use of the verb *divide* and our use of the verb *cleave*. Any division of a space results in *otherwise indistinguishable divisions of a state*, which are all at the same level, whereas a severance or cleavage shapes *distinguishable states*, which are at different levels.

An idea of the relative strengths of severance and division may be gathered from the fact that the rule of number is sufficient to unify a divided space, but not to void a cloven space.

Chapter 5

In eliciting rules for algebraic manipulation the text explicitly refers to the existence of systems of calculation other than the system described. This reference is both deliberate and inessential. It marks the level at which these systems are usually fitted out with their false, or truncated, or postulated, origins.

It is deliberate to inform the reader that, in the system of calculation we are building, we are not departing from the basic methods of other systems. Thus what we arrive at, in the end, will serve to elucidate them, as well as to fit them with their true origin. But, at the same time, it is important for the reader to see that the reference to other systems is inessential to the development of the argument in the text. For here it stands or falls on its own merit, dependent in no way for its validity upon agreement or disagreement with other systems. Thus rules 1 and 2, as can be seen from their justifications, say nothing that has not, in the text, already been said. They merely summarize the commands and instructions that will be relevant to the new kind of calculation we are about to undertake.

The replacement referred to in rule 2 is usually confined to independent variable expressions of simple (i.e. literal) form, and is in fact so confined in the text. But the greater licence granted by the rule is not devoid of significant application, if required.

Chapter 6

By the revelation and incorporation of its own origin, the primary algebra provides immediate access to the nature of the relationship between operators and operands. An operand in the algebra is merely a conjectured presence or absence of an operator.

This partial identity of operand and operator, which is not confined to Boolean algebras, can in fact be seen if we extend more familiar descriptions, although in these descriptions it is not so obvious. For example, we can find it by taking the Boolean operators $\vee$ (usually interpreted as the logical 'or', but here used purely mathematically) and . (usually interpreted as the logical 'and', but here again used purely mathematically), freeing their scope (as, by the principle of relevance, we may), freeing the order of the variables within their scope (as, by the same principle, we also may), and extrapolating mathematically to the case of no variable,

	...(a b c)	$\vee$.	(a b)	$\vee$.	(a)	$\vee$.	()	$\vee$.
permute	1 1 1	1 1	1 1	1 1	1	1 1		0 1
permute	1 1 0	1 0	1 0	1 0	0	0 0		
permute	1 0 0	1 0	0 0	0 0				
permute	0 0 0	0 0						

which shows quite plainly that we have no need of the arithmetical forms 0, 1 (or z, u, or F, T, etc), since we can equate them with ()$\vee$ and (). respectively. We can now write a Boolean variable of the form a, b, etc wherever we conjecture the presence of one of these two fundamental particles, but are not sure (or don't care) which. The functional tables for $\vee$ and . of two variables thus become

(a	b)	$\vee$	.
(()$\vee$	()$\vee$)	()$\vee$	()$\vee$
(()$\vee$	().)	().	()$\vee$
(().	().)	().	(). ,

the permutation being assumed.

J1, J2 are not the only two initials which may be taken to determine the primary algebra. We see[11] from Huntington's fourth postulate-set that we could have used C5, C6.

[11] Edward V Huntington, *Trans. Amer. Math. Soc.*, 35 (1933) 280–5.

The demonstration of J1, J2 from C5, C6 is both difficult and tedious. This is evidently because we find two basic algebraic principles, in one of which a variable is transplanted in the expression, and in the other of which it is eliminated from it. Provided we keep these two principles apart, subsequent demonstrations are not difficult. If, as in Huntington's two equations, they are inter-mingled, then their subsequent unravelling can be difficult.

Our expression here of Huntington's equations in the form of C5, C6 is not in the form in which he originally expressed them. He was hampered by the crippling assumptions of order relevance and binary scope, with which we have not at any stage weakened the primary algebra. For this reason he found it necessary to give two more equations to complete the set. C5 and C6, considered as initials, are of interest chiefly because they employ only two distinct variables, whereas J1 and J2 employ three.

I had at first supposed the demonstration of C1 to be impossible from J1 and J2 as they stand. In 1965 a pupil, Mr John Dawes, produced a rather long proof to the contrary, so the following year I set the problem to my class as an exercise, and was rewarded with a most elegant demonstration by Mr D A Utting. I use Mr Utting's demonstration, slightly modified, in the text.

Although, superficially, it may look less efficient, it is, eventually, more natural and convenient to use names rather than numbers to identify the more important consequences, as indeed it is with theorems, since they do not in general form an ordered set.

In naming such consequences I have aimed to find what seems appropriate as a description of the named process, as it appears in the algebra, without doing violence to its arithmetical origin. In some places both the forms and the names are recognizably similar to those of other authors who have determined Boolean algebras. In most such cases hitherto, the commonly used name describes only one of the directions in which the step can be taken. What is called Boolean expansion is an example. In such a case, where the name is appropriate to the step as taken in one direction only, I have introduced an antonym for the other direction, and given a generic name to cover both. In other recognizable cases I have found what seems to me to be a more appropriate name, such as occultation for what Whitehead called[12] absorption. The occulting part of the expression is not so much absorbed in the remainder as eclipsed by it. This can be seen quite plainly in the arithmetic, or alternatively if the expression is illustrated with a Venn diagram. To the best of my knowledge, Peirce was the only previous author to recognize, as such, what I call position. He called[13] it erasure, thus again drawing attention to only one direction of application.

I do not suppose all the names will always stick. Familiarity tends to produce a kind of in-slang, often more appropriate, in its place, than what is deemed to be academically proper or seemly. For example, the engineering application of consequence 2 has produced the more homely 'breed' for 'regenerate', and 'revert' for 'degenerate', and it is of interest to note that the transformations of this consequence are immediate images of what Proclus called[14] πρόοδος and ἐπιστροφή, translated by Dodds into *procession* and *reversion*.

The fact that descriptive names such as 'transposition' and 'integration' are differently applied elsewhere in mathematics (and, indeed, elsewhere in this book) does not appear to be a reason for avoiding their use in the senses defined in this chapter. The deeper the level of investigation, the harder it becomes to find words strong enough to cover what is found there, and in all cases my use of language to describe primitive processes draws on a greater power of signification than is needed for its more superficial and specialized uses.

One of the most beautiful facts emerging from mathematical

[12] Alfred North Whitehead, *A treatise on universal algebra*, Vol. I, Cambridge, 1898, p 36.

[13] Charles Sanders Peirce, *Collected papers*, Vol. IV, Cambridge, Massachusetts, 1933, pp 13–18.

[14] *ΠΡΟΚΛΟΥ ΔΙΑΔΟΧΟΥ ΣΤΟΙΧΕΙΩΣΙΣ ΘΕΟΛΟΓΙΚΗ* with a translation by E R Dodds, 2nd edition, Oxford, 1963.

studies is this very potent relationship between the mathematical process and ordinary language. There seems to be no mathematical idea of any importance or profundity that is not mirrored, with an almost uncanny accuracy, in the common use of words, and this appears especially true when we consider words in their original, and sometimes long forgotten, senses.

The fact that a word may have different, but related, meanings at different, but related, levels of consideration does not normally render communication impossible. On the contrary, it is evident that communication of any but the most trivial ideas would be impossible without it.

Since at this point in the text the fundamental forms of mathematical communication are now practically complete, it may be a revealing exercise to retranslate into longhand some of the shorthand forms developed by application of the canon of contracting reference. For this purpose we take the statement and demonstration of consequence 9 (p 35). In words and figures it could run thus.

The ninth consequence, called crosstransposition, or C9 for short, may be stated as follows.

b cross *r* cross cross all *a*
cross *r* cross cross 2 *x* cross
r cross 2 *y* cross *r* cross 2
cross all

expresses the same value as

r cross *ab* cross all *rxy* cross 3.

When the step allowed by this equation is taken from the former to the latter expression, it is called to crosstranspose or collect, and when taken in reverse it is called to crosstranspose or distribute.

The equation can be demonstrated thus.

b cross *r* cross cross all *a*
cross *r* cross cross 2 *x* cross
r cross 2 *y* cross *r* cross 2
cross all

may be changed to

> *b* cross *r* cross cross all *a*
> cross *r* cross cross 2 *xy*
> cross 2 *r* cross 2 cross all

by using C1, J2, and then C1 again. This in turn may be changed to

> *baxy* cross 2 *r* cross 2 cross
> all *rxy* cross 2 *r* cross 2
> cross 2

by C8 and then by applying C1 three times, etc.

We may observe that, in expressions, the mathematical language has become entirely visual, there is no proper spoken form, so that in reverbalizing it we must *encode* it in a form suitable for ordinary speech. Thus, although the mathematical form of an expression is clear, the reverbalized form is obscure.

The main difficulty in translating from the written to the verbal form comes from the fact that in mathematical writing we are free to mark the two dimensions of the plane, whereas in speech we can mark only the one dimension of time.

Much that is unnecessary and obstructive in mathematics today appears to be vestigial of this limitation of the spoken word. For example, in ordinary speech, to avoid direct reference to a plurality of dimensions, we have to fix the scope of constants such as 'and' and 'or', and this we can most conveniently do at the level of the first plural number. But to carry the fixation over into the written form is to fail to realize the freedom offered by an added dimension. This in turn can lead us to suppose that the binary scope of operators assumed for the convenience of representing them in one dimension is something of relevance to the actual form of their operation, which, in the case of simple operators even at the verbal level, it is not.

Chapter 7

In the description of theorem 14 'the constant' refers to the operative constant. There are two constants in the calculus, a mark or operator, and a blank or void. Reference to 'the constant' without qualification will usually be taken to denote the operator rather than the void.

Chapter 8

We have already distinguished, in the text, between demonstration and proof. In making this distinction, which appears quite natural, we see at once that a proof can never be justified in the same way as a demonstration. Whereas in a demonstration we can see that the instructions already recorded are properly obeyed, we cannot avail ourselves of this procedure in the case of a proof.

In a proof we are dealing in terms which are outside of the calculus, and thus not amenable to its instructions. In any attempt to render such proofs themselves subject to instruction, we succeed only at the cost of making another calculus, inside of which the original calculus is cradled, and outside of which we shall again see forms which are amenable to proof but not demonstration.

The validity of a proof thus rests not in our common motivation by a set of instructions, but in our common experience of a state of affairs. This experience usually includes the ability to reason which has been formalized in logic, but is not confined to it. Nearly all proofs, whether about a system containing numbers or not, use the common ability to compute, i.e. to count* in either direction, and ideas stemming from our experience of this ability.

It seems open to question why we regard the proof of a theorem as amounting to the same degree of certainty as the demonstration of a consequence. It is not a question which, at first sight, admits of an easy answer. If an answer is possible, it would seem to lie in the concept of experience. We gain experience of living representative processes, in particular of

* Although *count* rests on *putare* = prune, correct, (and hence) reckon, the word *reason* comes from *reri* = count, calculate, reckon. Thus the reasoning and computing activities of proof were originally considered as one. We may note further that *argue* is based on *arguere* = clarify (literally 'make silver'). We thus find a whole constellation of words to do with the process of *getting it right*.

argument and of counting forwards and backwards in units, and through this experience become quite certain, in our own minds, of the validity of using it to substantiate a proof. But since the procedures of the proof are not, themselves, yet codified in a calculus (although they may eventually become so), our certainty at this stage must be deemed to be intuitive. We can achieve a demonstration simply by following instructions, although we may be unfamiliar with the system in which the instructions are obeyed. But in proving a theorem, if we have not already codified the structure of the proof *in* the form of a calculus, we must at least be familiar with, or experienced in, whatever it is we take to be the *ground* of the proof, otherwise we shall not *see* it *as* a proof.

Another way of regarding the relationship between demonstration and proof, which adds support to the proposition that the degree of certainty of a proof is equal to that of a demonstration, is to consider it as the boundary dividing the state of proof from the state of demonstration. A demonstration, we remember, occurs inside the calculus, a proof outside. The boundary between them is thus a shared boundary, and is what is approached, in one or the other direction, according to whether we are demonstrating a consequence or proving a theorem. Thus consequences and theorems can be seen to bear to each other a fitting relationship.

But the boundary marking their relationship, although shared, is (like the existential boundary (see pp 124 sq)) seen from one side only, since if we know the ground on which a demonstration rests (i.e. provided we understand the formal, as distinct from the pragmatic, reasons for the initial equations we employ, and so do not have to postulate them), the demonstration can be seen as a proof by implication, although a proof is never seen as a demonstration. We observe, in fact, that demonstration bears the same relationship to proof as initial equation bears to axiom, but we should also note that the relationship is evident for arithmetic only, and is lost when we make the departure into algebra. This appears to be why algebras are commonly presented without axioms, in any proper sense of the word.

The fact that a proof is a way of making apparently obvious

what was already latently so is of some mathematical interest. Although there are any number of distinct proofs of a given theorem, they can all, even so, be hard to find. In other words, we can set about trying to prove a theorem in a large number of wrong ways before coming across a right way.

Even the analogy of seeking something cannot, in this context, be quite right. For what we find, eventually, is something we have known, and may well have been consciously aware of, all along. Thus we are not, in this sense, seeking something that has ever been hidden. The idea of performing a search can be unhelpful, or even positively obstructive, since searches are in general organized to find something which has been previously hidden, and is thus not open to view.

In discovering a proof, we must do something more subtle than search. We must come to see the *relevance*, in respect of whatever statement it is we wish to justify, of some fact in full view, and of which, therefore, we are already constantly aware. Whereas we may know how to undertake a search for something we can *not* see, the subtlety of the technique of trying to 'find' something which we already *can* see may more easily escape our efforts.

This might be a helpful moment to introduce a distinction between following a course of argument and understanding it. I take understanding to be the experience of what is understood in a wider context. In this sense, we do not fully understand a theorem until we are able to contain it in a more general theorem. We can nevertheless follow its proof, in the sense of coming to see its evidence, without understanding it in the wider sense in which it may rest.

Following and understanding, like demonstrating and proving, are sometimes wrongly taken as synonymous. Very often a person is regarded as not understanding an argument, a process, a doctrine, when all that is certain is that he has not followed it. But his failure to follow may be quite deliberate, and may arise from the fact that he *has* understood what was presented to him, and does not follow it because he sees a shorter, or otherwise more acceptable, path, although he might not, yet, know how to communicate it.

Following may thus be associated particularly with doctrine, and doctrine demands an adherence to a particular way of saying or doing something. Understanding has to do with the fact that what ever is said or done can always be said or done a different way, and yet all ways remain the same.

Chapter 9

We observe that the idea of completeness cannot apply to a calculus as a whole, but only to a representation of one determination of it by another. What is questioned, in fact, is the completeness of an alternative form of expression.

The paragon of such an alternative is the algebraic representation of an arithmetic, although we do in fact find a more central case of it in the arithmetical representation of a form. In the latter case, as we see from the theorems of representation, the idea of completeness condenses with that of consistency. In the less central case, the two ideas come apart. Thus the most primitive example of completeness, in its pure form, is to be found in algebraic representation.

A fact to which Gödel drew attention [5] is that an algebra which includes representations of addition and multiplication *cannot* fully account for an arithmetic of the natural numbers in which these operations are taken as elementary. Thus, in number theory, although certain relationships can be proved, no algebra can be constructed in which all such relationships are demonstrable.

The advent of Gödel's theorem has never seemed to me to be a reason for despair, as some investigators have taken it to be, but rather an occasion for celebration, since it confirms what men of mathematics have found from experience, notably that ordinary arithmetic is a richer ground for investigation than ordinary algebra.

Chapter 10

It is usual to prove the independence of initial equations indirectly[15]. It is not commonly observed, although it becomes

[15] following Edward V Huntington, *Trans. Amer. Math. Soc.*, 5 (1904) 288–309.

evident when we consider it, that with a set of only two initials, a direct proof of their independence is always available, and I give such a proof in the text.

An independence proof may be properly considered as an incompleteness proof of the calculus with the missing initial.

Chapter 11

The question of whether or not functions of themselves are allowable has been discussed at wearisome length by many authorities [cf 8] since *Principia mathematica* was published. The Whitehead-Russell argument for disallowing them is well known. It is the subject of a number of comments by Wittgenstein [4, propositions 5.241 sq]. (I use the Pears-McGuinness translation for what follows.)

An operation, says Wittgenstein, is not the mark of a form, but of a relation between forms. Wittgenstein here sees what I call the mark of distinction between states, which he calls forms, and also sees its connexion with the idea of operation. He then remarks [5.251] that

> a function cannot be its own argument, whereas an operation can take one of its own results as its base.

This applies only, in the strict sense, to single-valued functions. If we allow inverse and implicit functions, then the assertion above is untrue. A function of a variable, in the wider meaning with which it is defined in this chapter, is the result of a possible set of operations on the variable. Thus if an operation can take its own result as a base, the function determined by this operation can be its own argument.

I shall proceed, in the light of this relaxation, to examine in some detail the analogy between Boolean equations and those of an ordinary numerical algebra.

Boole maintained[16] that the equation with which he defined what he called the law of duality, notably

$$x^2 = x,$$

[16] George Boole, *An investigation of the laws of thought*, Cambridge, 1854, p 50.

is of the second degree. So it is, as stated, but by it he determines that, in his notation, all equations of degree >1 shall be reduced to the first degree. In other words, it is an equation of the second degree only at the descriptive level, not in the algebra itself.

The spuriousness of its alleged degree, considered in the algebra itself, is revealed by Boole's assertion in a footnote [p 50] that an equation of the third degree has no interpretation in his algebra. It has, as we shall presently see, but Boole appears at this point to have been overcome by his notation, which uses numerical forms for an algebra which is essentially non-numerical.

Boole's equation

$$x^2 = x$$

is an analogue, in the primary algebra, of

$$aa = a.$$

This, as we see, is an equation of the first degree, being expressible without subversion. The real form of the analogy with a numerical algebra may be illustrated as follows.

Suppose

$$px^2 + qx + r = 0$$

where p, q, r may stand for rational numbers. We can re-express this equation in the form

F1 $$x^2 + ax + b = 0$$

by calling $q/p = a$ and $r/p = b$, and it may then be further transposed into

$$x = -a + \frac{-b}{x}$$

F2 $$= -a + \cfrac{-b}{-a + \cfrac{-b}{-a + \cfrac{-b}{\cdots}}}.$$

In a Boolean algebra we are properly denied the mode of F1, but permitted the mode of F2, which is either continuous or, if we want to see it so, subversive. Thus an equation of any degree is both constructible and meaningful in a Boolean algebra, although not necessarily in the primary form of it. To reach a higher degree, all we need to do is to add a distinct subversion. The two modulator equations at the end of the chapter are both of degree >2. They were first developed in 1961, in collaboration with Mr D J Spencer Brown, for special-purpose computer circuits. Such equations undertake an excursion to a higher order of infinity, and, although still expressible in subversive form, they cannot be represented in continuous form on a plane.

The circuits represented by these equations, the latter being presently in use by British Railways, comprise, as far as we know, a first application of each of two inventions, notably the first construction of a device which counts entirely by 'logic' (i.e. with switches only, and with no artificial time delays such as electrical condensers) and, in addition, the first use, in a switching circuit, of imaginary Boolean values in the course of the construction of a real answer. This latter might in fact be the first use of such imaginary values for any purpose, although it is my guess that Fermat (who was apparently too excellent a mathematician to make a false claim to a proof) used them in the proof of his great theorem, hence the 'truly remarkable' nature of his proof, as well as its length.

The fact that imaginary values *can* be used to reason towards a real and certain answer, coupled with the fact that they *are not* so used in mathematical reasoning today, and also coupled with the fact that certain equations plainly *cannot* be solved without the use of imaginary values, means that *there* **must** *be mathematical statements* (whose truth or untruth is in fact perfectly decidable) *which cannot be decided by the methods of reasoning to which we have hitherto restricted ourselves.*

Generally speaking, if we confine our reasoning to an interpretation of Boolean equations of the first degree only, we should *expect* to find theorems which will always defy decision, and the fact that we do seem to find such theorems in common arithmetic may serve, here, as a practical confirmation of this obvious prediction. To confirm it theoretically, we need only to prove (1) that such theorems *cannot* be decided by reasoning of the first degree, and (2) that they *can* be decided by reasoning of a higher degree. (2) would of course be proved by providing such a proof of one of these theorems.

I may say that I believe that at least one such theorem will shortly be decided by the methods outlined in the text. In other words, I believe that I have reduced their decision to a technical problem which is well within the capacity of an ordinary mathematician who is prepared, and who has the patronage or other means, to undertake the labour.

Any evenly subverted equation of the second degree might be called, alternatively, evenly informed. We can see it over a sub-version (turning under) of the surface upon which it is written, or alternatively, as an in-formation (formation within) of what it expresses.

Such an expression is thus informed in the sense of having its own form within it, and at the same time informed in the sense of remembering what has happened to it in the past.

We need not suppose that this is exactly how memory happens in an animal, but there are certainly memories, so-called, constructed this way in electronic computers, and engineers have constructed such in-formed memories with magnetic relays for the greater part of the present century.

We may perhaps look upon such memory, in this simplified in-formation, as a precursor of the more complicated and varied forms of memory and information in man and the higher animals. We can also regard other manifestions of the classical forms of physical or biological science in the same spirit.

Thus we do not imagine the wave train emitted by an excited finite echelon to be exactly like the wave train emitted from an excited physical particle. For one thing the wave form from an echelon is square, and for another it is emitted without energy. (We should need, I guess, to make at least one more departure from the form before arriving at a conception of energy on these lines.) What we see in the forms of expression at this stage, although recognizable, might be considered as simplified precursors of what we take, in physical science, to be the real thing. Even so, their accuracy and coverage is striking. For example, if, instead of considering the wave train emitted by the expression in Figure 4, we consider the expression itself, in its quiescent state, we see that it is composed of standing waves. If, therefore, we shoot such an expression through its own representative space, it will, upon passing a given point, be observable at that point as a simple oscillation with a frequency proportional to the velocity of its passage. We have thus already arrived, even at this stage, at a remarkable and striking precursor of the wave properties of material particles.

We may look upon such manifestations as the formal seeds, the existential forerunners, of what must, in a less central state, under less certain conditions, come about. There is a tendency, especially today, to regard existence as the source of reality, and thus as a central concept. But as soon as it is formally examined (cf Appendix 2), existence* is seen to be highly peripheral and, as such, especially corrupt (in the formal sense) and vulnerable. The concept of truth is more central, although still recognizably peripheral. If the weakness of present-day science is that it centres round existence, the weakness of present-day logic is that it centres round truth.

Throughout the essay, we find no need of the concept of truth, apart from two avoidable appearances (true = open to proof) in the descriptive context. At no point, to say the least, is it a necessary inhabitant of the calculating forms. These forms are thus not only precursors of existence, they are also precursors of truth.

It is, I am afraid, the intellectual block which most of us come up against at the points where, to experience the world clearly, we must abandon existence to truth, truth to indication, indication to form, and form to void, that has so held up the development of logic and its mathematics.

What status, then, does logic bear in relation with mathematics? We may anticipate, for a moment, Appendix 2, from

* *ex* = out, *stare* = stand. Thus to exist may be considered as to stand outside, to be exiled.

which we see that the arguments we used to justify the calculating forms (e.g. in the proofs of theorems) *can themselves be justified by putting them in the form of the calculus*. The process of justification can be thus seen to feed upon itself, and this may comprise the strongest reason against believing that the codification of a proof procedure lends evidential support to the proofs in it. All it does is provide them with coherence. A theorem is no more proved by logic and computation than a sonnet is written by grammar and rhetoric, or than a sonata is composed by harmony and counterpoint, or a picture painted by balance and perspective. Logic and computation, grammar and rhetoric, harmony and counterpoint, balance and perspective, can be seen in the work *after* it is created, but these forms are, in the final analysis, parasitic on, they have no existence apart from, the creativity of the work itself. Thus the relation of logic to mathematics is seen to be that of an applied science to its pure ground, and all applied science is seen as drawing sustenance from a process of creation with which it can combine to give structure, but which it cannot appropriate.

Chapter 12

Let us imagine that, instead of writing on a plane surface, we are writing on the surface of the Earth. Ignoring rabbit holes, etc, we may take it to be a surface of genus 0. Suppose we write

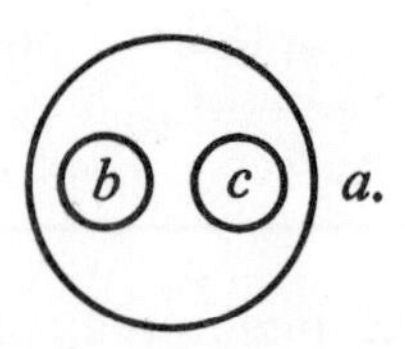

To make it readable from another planet, we write it large. Suppose we draw the outer bracket round the Equator, and make the brackets containing b and c follow the coastlines of Australia and the South Island of New Zealand respectively.

Above is how the expression will appear from somewhere in the Northern Hemisphere, say London. But let us travel.

Arriving at Cape Town we see

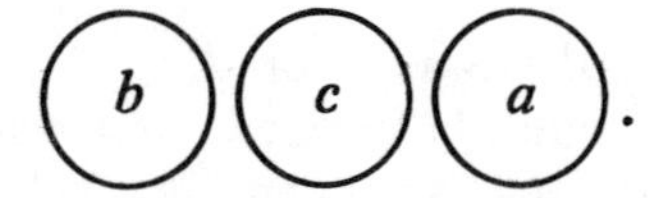

Sailing on to Melbourne, we see

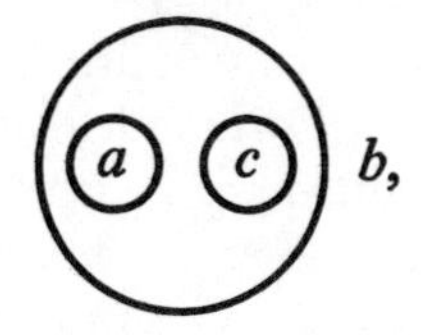

and proceeding from there to Christchurch, we see

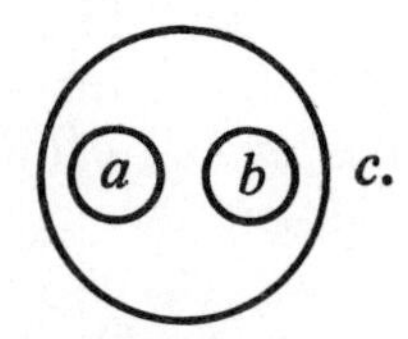

These four expressions are distinct and not equivalent. Thus it is evidently not enough merely to write down an expression, even on a surface of genus 0, and expect it to be understood. We must also indicate where the observer is supposed to be standing in relation to the expression. Writing on a plane, the ambiguity is not apparent because we tend to see the expression from outside of the outermost bracket. When it is written on the surface of a sphere, there may be no means of telling which of the brackets is supposed to be outermost. In such a case, to make an expression meaningful, we must add to it an indicator to present a place from which the observer is invited to regard it.

We observe in the third experiment an alternative way (although here less powerful) of using the principle of relevance. By the normal use of the principle we could obliterate the additional markings (since every state is identically marked) and arrive at the single circle in one step, whereas in the experiment we take the weaker course of obliterating the line of distinction between the markings, and then need one more step to reach the single circle.

Note that both of these ways of simplification are *different* from the methods of cancellation and condensation adopted for the calculus, although arising from, and thus not inconsistent with, them. From the experiment we begin to see in fact how all the constellar principles by which we navigate our journeys out from and in to the form spring from the ultimate reducibility of numbers and voidability of relations. It is only by arresting or fixing the use of these principles at some stage that we manage to maintain a universe in any form at all, and our understanding of such a universe comes not from discovering its present appearance, but in remembering what we originally did to bring it about.

In this way the calculus itself can be realized as a direct recollection. As we left the central state of the form, proceeding outwards and imagewise towards the peripheral condition of existence, we saw how the laws of calling and crossing, which set the stage of our journey through representative space, became fixed stars in the familiar play of time. Our projected hopes and fears of their ultimate atonement, which we called theorems, became their supporting cast. In the end, as we re-enter the form, they are all justified and expended. They were needed only as long as they were doubted. When they cannot be doubted, they can be discarded.

Returning, briefly, to the idea of existential precursors, we see that if we accept their form as endogenous to the less primitive structure identified, in present-day science, with reality, we cannot escape the inference that what is commonly now regarded as real consists, in its very presence, merely of tokens or expressions. And since tokens or expressions are considered to be *of* some (other) substratum, so the universe itself, as we know it, may be considered to be an expression of a reality other than itself.

Let us then consider, for a moment, the world as described by the physicist. It consists of a number of fundamental particles which, if shot through their own space, appear as waves, and are thus (as in Chapter 11), of the same laminated structure as pearls or onions, and other wave forms called electromagnetic which it is convenient, by Occam's razor, to consider as travelling through space with a standard velocity. All these appear bound by certain natural laws which indicate the form of their relationship.

Now the physicist himself, who describes all this, is, in his own account, himself constructed of it. He is, in short, made of a conglomeration of the very particulars he describes, no more, no less, bound together by and obeying such general laws as he himself has managed to find and to record.

Thus we cannot escape the fact that the world we know is constructed in order (and thus in such a way as to be able) to see itself.

This is indeed amazing.

Not so much in view of what it sees, although this may appear fantastic enough, but in respect of the fact that it *can* see *at all.*

But *in order* to do so, evidently it must first cut itself up into at least one state which sees, and at least one other state which is seen. In this severed and mutilated condition, whatever it sees is *only partially* itself. We may take it that the world undoubtedly is itself (i.e. is indistinct from itself), but, in any attempt to see itself as an object, it must, equally undoubtedly, act* so as to make itself distinct from, and therefore false to, itself. In this condition it will always partially elude itself.

It seems hard to find an acceptable answer to the question of how or why the world conceives a desire, and discovers an ability, to see itself, and appears to suffer the process. That it does so is sometimes called the original mystery. Perhaps, in view of *the form* in which *we* presently *take* ourselves *to exist*, the mystery *arises from* our insistence on *framing* a question where there is, in reality, *nothing* to question. However it may appear, if such desire, ability, and sufferance be granted, the state or condition that arises as an outcome is, according

* Cf ἀγωνιστής = actor, antagonist. We may note the identity of action with agony.

to the laws here formulated, absolutely unavoidable. In this respect, at least, there is no mystery. We, as universal representatives, *can* record universal law far enough to say

> and so on, and so on you will eventually construct the universe, in every detail and potentiality, as you know it now; but then, again, what you will construct will not be all, for by the time you will have reached what now is, the universe will have expanded into a new order to contain what will then be.

In this sense, in respect of its own information, the universe *must* expand to escape the telescopes through which we, who are it, are trying to capture it, which is us. The snake eats itself, the dog chases its tail.

Thus the world, when ever it appears as a physical universe*, must always seem to us, its representatives, to be playing a kind of hide-and-seek with itself. What is revealed will be concealed, but what is concealed will again be revealed. And since we ourselves represent it, this occultation will be apparent in our life in general, and in our mathematics in particular. What I try to show, in the final chapter, is the fact that we really knew all along that the two axioms by which we set our course were mutually permissive and agreeable. At a certain stage in the argument, we somehow cleverly obscured this knowledge from ourselves, in order that we might then navigate ourselves through a journey of rediscovery, consisting in a series of justifications and proofs with the purpose of again rendering, to ourselves, irrefutable evidence of what we already knew.

Coming across it thus again, in the light of what we had to do to render it acceptable, we see that our journey was, in its preconception, unnecessary, although its formal course, once we had set out upon it, was inevitable.

* *unus* = one, *vertere* = turn. Any given (or captivated) universe is what is seen as the result of a making of one turn, and thus *is the appearance* of any first distinction, and only a minor aspect of all being, apparent and non-apparent. Its particularity is the price we pay for its visibility.

APPENDIX 1

PROOFS OF SHEFFER'S POSTULATES

Sheffer's postulates [3, p 482] for Boolean algebras are chosen for proof because they comprise, amongst those that are widely known, the least such set. They do not constitute, under the constraints he adopted, the least possible such set*.

Sheffer's description, quoted below, is in fact complete (although not proved to be so at the time), so that proofs of the postulates in it will serve to prove all postulates in every description of Boolean algebra. None, as far as I know, has been proved before.

He assumes

I. A class K,

II. A binary K-rule of combination $|$,

III. The following properties of K and $|$:

1. There are at least two distinct K-elements.

2. Whenever a and b are K-elements, $a|b$ is a K-element.

   Def. $a' = a|a$.

3. Whenever a and the indicated combinations of a are K-elements,

$$(a')' = a.$$

4. Whenever a, b, and the indicated combinations of a and b are K-elements,

$$a|(b|b') = a'.$$

* For example, Sheffer's three initial equations can be derived from the two initials $a'|(b'|b) = a$ and $((c'|a)|(b'|a))' = a|(b|c)$.

5. Whenever a, b, c, and the indicated combinations of a, b, and c are K-elements,

$$(a|(b|c))' = (b'|a)|(c'|a).$$

We aim to prove each of the propositions numbered 1–5.

Proofs

1. Let the class K be the set of indicators of the states distinguished by the first distinction. There are two such states, from which the first proposition follows.

2. Let $a|b$ be written for $\overline{ab}\rceil$. The second proposition evidently follows.

Let a' be written for $\overline{a}\rceil$. Sheffer's definition

$$a' = a|a$$

follows since

$$\overline{a}\rceil = \overline{aa}\rceil \qquad \text{C5.}$$

Now, if each literal variable is a K-element,

3. $$\overline{\overline{a}\rceil}\rceil = a \qquad \text{C1}$$

may be written

$$(a')' = a,$$

4. $$\overline{a\ \overline{b\ \overline{b}\rceil}\rceil}\rceil = \overline{a}\rceil \qquad \text{J1}$$

may be written

$$a|(b|b') = a',$$

and

5. $$\overline{\overline{a\ \overline{b\ c}\rceil}\rceil}\rceil = \overline{\overline{\overline{\overline{b}\rceil\ a}\rceil\ \overline{\overline{c}\rceil\ a}\rceil}\rceil}\rceil \qquad \text{C1 (thrice), J2}$$

may be written

$$(a|(b|c))' = (b'|a)|(c'|a).$$

This accounts for the third, fourth, and fifth propositions, and completes the proofs.

Note 1. By the principle of relevance, the stroke in Sheffer's notation may be omitted. Proof of this, which is left with the reader, is perhaps somewhat harder than the immediate apprehension of its truth.

Note 2. Sheffer explicitly assumes the restriction of his operator to a binary scope, and also, implicitly, assumes the relevance of the order in which the variables under operation appear. Each of these assumptions is in fact less central to mathematics than is commonly supposed, and neither is necessary at this stage. Sheffer was therefore forced to design his initial equations so ingeniously as to contradict them both. The latter he can contradict explicitly, without the disorder becoming too apparent, by allowing $a|b = b|a$ as a consequence, but he cannot explicitly contradict the former without obviously denying a rule already recorded, and this would appear foolish, although it is, in fact, now the best way out of the deep trouble that such an ill-considered rule brings in its train. By allowing it to stand, Sheffer's description is rendered practically useless as a calculus.

To understand why Sheffer did not see this, let us take the unusual course of considering his position in the light of the social forces at work around him.

Discoveries of any great moment in mathematics and other disciplines, once they are discovered, are seen to be extremely simple and obvious, and make everybody, including their discoverer, appear foolish for not having discovered them before. It is all too often forgotten that the ancient symbol for the prenascence of the world* is a fool, and that foolishness, being

* *wer* = man, *ald* = age, old. The world may be taken to be the *manifest* properties of the all, its identity with the age of man being evident through the fact that man is a primary animal with a hand ('manifest' coming from *manus* = hand, *festus* = struck). Thus the world is considerably less than the all, which includes the unmanifest, but considerably greater than 'the' universe (more correctly, than *any* universe), which is merely the *formal appearance* of *one* of the possible manifestations which make up the world.

a divine state, is not a condition to be either proud or ashamed of.

Unfortunately we find systems of education today which have departed so far from the plain truth, that they now teach us to be proud of what we know and ashamed of ignorance. This is doubly corrupt. It is corrupt not only because pride is in itself a mortal sin, but also because to teach pride in knowledge is to put up an effective barrier against any advance upon what is already known, since it makes one ashamed to look beyond the bonds imposed by one's ignorance.

To any person prepared to enter with respect into the realm of his great and universal ignorance, the secrets of being will eventually unfold, and they will do so in a measure according to his freedom from natural and indoctrinated shame in his respect of their revelation.

In the face of the strong, and indeed violent, social pressures against it, few people have been prepared to take this simple and satisfying course towards sanity. And in a society where a prominent psychiatrist can advertise that, given the chance, he would have treated Newton to electric shock therapy, who can blame any person for being afraid to do so?

To arrive at the simplest truth, as Newton knew and practised, requires *years* of *contemplation*. Not activity. Not reasoning. Not calculating. Not busy behaviour of any kind. Not reading. Not talking. Not making an effort. Not thinking. Simply *bearing in mind* what it is one needs to know. And yet those with the courage to tread this path to real discovery are not only offered practically no guidance on how to do so, they are actively discouraged and have to set about it in secret, pretending meanwhile to be diligently engaged in the frantic diversions and to conform with the deadening personal opinions which are being continually thrust upon them.

In these circumstances, the discoveries that any person is able to undertake represent the places where, in the face of induced psychosis, he has, by his own faltering and unaided efforts, returned to sanity. Painfully, and even dangerously, maybe. But nonetheless returned, however furtively.

We may note in this connexion that Peirce [13], who discovered, some thirty years ahead of Sheffer, that the logic of propositions could be done with one constant, did not publish this discovery, although its importance must have been evident to him; that Stamm, who himself discovered and published[17] this fact two years before Sheffer, omits, in his paper, to make a simple and obvious substitution which would have put his claim beyond doubt; and that Sheffer [3], who ignores Stamm's paper, is currently credited with the major discovery recorded in it.

[17] E Stamm, *Monatshefte für Mathematik und Physik*, 22 (1911) **137–40.**

APPENDIX 2

THE CALCULUS INTERPRETED FOR LOGIC

The calculus of indications consists of a set of ways of indicating one or the other of the two states distinguished by the first distinction, so we shall be able to find an application of it to the indicative forms of any clear distinction of this kind. It must, for example, apply to cases where doors can be open or shut, or where switches can be on or off, or where lines can be clear or blocked. It will also apply to a language structure in which sentences can be true or false.

Considering the question of its application in the light of the direction from which we have come, it is not immediately obvious that the calculus will have a *useful* or *revealing* application to any of these cases, even though we can see it will apply. The calculus has been built up, in the essay, in a series of forms and departures, and although what we have found there may seem curious, why we took the trouble to look for it may seem equally so.

The fact is, in undertaking the development of the calculus in this direction, the author is making the journey a second time, whereas it may be the first journey for his reader. The author's previous journey was in the opposite direction, from the forms of interpretation we are now about to discuss, towards the form of indication from which they arise. So he is aware, although his reader may not yet be, of how and where it will end, and of the clarifications and simplifications he had to undertake in order to find the way to the place from which he is now returning. He knows, also, that these clarifications will become strengthened on the return journey, although he may still have to convey his vision of their clarity and impression of their strength to the reader.

In interpreting a calculus, what we do is match the values or states or elements allowed in the calculus to a similar set of values or states or elements in what is to become its interpretation. An interpretation is properly matched if each element in it is associated with an identifiable element in the calculus, and the elements in each case have similar distinctions between them. Even so, although there must be this degree of similarity between a calculus and an interpretation of it, in any case of a calculus of more than one value, the calculus and the interpretation are distinct. The fact of their distinction is made plain by the plurality of ways in which a given interpretation can be applied.

With a calculus representing n distinct values there are evidently $n!$ different ways of matching them with n distinct values represented in the interpretation, and thus $n!$ different forms that such an interpretation can take. In interpreting the calculus of indications for sentential logic, we shall match one each of the states of the primary distinction with one each of the states distinguished by what is true and what is not true, which will offer us $2! = 2$ possible interpretative choices.

The fact that a calculus and an interpretation of it are distinct entities is of crucial importance. By failure to make use of it, we cut ourselves off from forms of simplification which are otherwise readily available. One such form, recognizably frequent in mathematics, consists in the underlying use, when required, of a construction which is devoid of interpretation within the particular application, but which can nevertheless be used to shorten the way to an answer there. A notable example, from outside the field of logic, is the use, as an operator, of $i = \sqrt{-1}$ in electromagnetic theory.

We see, in logic, that 'not true' means the same as 'false', and that 'not false' also means 'true'. So we have a choice of whether to associate the unmarked state with truth and the marked state with untruth, or to associate the marked state with truth and the unmarked state with untruth. Although it is quite immaterial, from the point of view of calculation, which we do, the latter arrangement is in fact easier from the point of view of interpretation.

Accordingly, we identify the marked state, and thereby an

empty cross, with true, and the unmarked state, and therein a blank space, with false.

We can now let variables $a, b, \ldots$ stand for the possible truth values of the various simple sentences in a complex sentence, and for this purpose we may allot a distinct variable to each distinct simple sentence.

Next we must find forms, in the primary algebra, which will properly represent the constants, in the sentential calculus, by which these values are related.

It is clear that we can interpret $\sim a$, or *not a*, through $\overline{a|}$. It is also clear that a truth table for $a \vee b$, or *a and/or b*, has exactly the same form as that displayed by the rule of dominance, so that $a \vee b$ can be represented simply by *ab*. All other forms can now be built up from these. Thus

in words	*in the sentential calculus*	*in the primary algebra*
not a	$\sim a$	$\overline{a|}$
a or b	$a \vee b$	ab
a and b	$a \, . \, b$	$\overline{\overline{a|}\ \overline{b|}\,|}$
a implies b	$a \supset b$	$\overline{a|}\ b$.

It is the simplicity, in this interpretative choice, of the representation of implication which renders it easier than the alternative,

in which $a \supset b$ must be written $\overline{\overline{b|}\ a\,|}$.

In examining the interpretation as thus set out, we at once see two sources of power which are both unavailable to the standard sentential calculus. They are, notably, the condensation of a number of representative forms into one form, and the ability to proceed, where required, beyond logic through the primary arithmetic.

Regarding the first of these sources, we may take, for the purpose of illustration, the forms for logical conjunction. In the sentential calculus they are

$$a \,.\, b$$
$$b \,.\, a$$
$$\sim(\sim a \vee \sim b)$$
$$\sim(\sim b \vee \sim a)$$
$$\sim(a \supset \sim b)$$
$$\sim(b \supset \sim a).$$

Each of these six distinct expressions is written, in the primary algebra, in only one way,

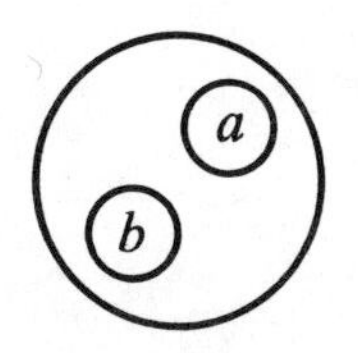

This is a proper simplification, since the object of making such sentences correspond with these symbols is not representation, but calculation. Thus, by the mere principle of avoiding an unnecessary prolixity in the representative form, we make the process of calculation considerably less troublesome.

But the power granted to us through this simplicity, although great, is itself small compared with the power available through the connexion of the primary algebra with its arithmetic. For this faculty enables us to dispense with a whole set of lengthy and tedious calculations, and also with their no less troublesome alternatives, such as the exhaustive (and mathematically weak) procedures of truth tabulation, and the graphical (and thus mathematically unsophisticated) methods of Venn diagrams and their modern equivalents.

This is made possible by the fact that the three classes of algebraic expression, integral, disintegral, and consequential, which correspond, in the interpretation, with true (tautologous), false (contradictory), and contingent, are readily distinguishable by manipulation.

Example. Classify the following complex sentences in respect of their truth, untruth, or contingency.

1. $(q \supset r) \supset ((p \vee q) \supset (r \vee p))$.
2. $((r \supset p) \supset (\sim(p \vee q))) \,.\, p$.
3. $((p \supset q) \,.\, (r \supset s) \,.\, (q \vee s)) \supset (p \vee r)$.

$1 = \overline{\overline{q}\rceil\, r}\rceil\;\overline{pq}\rceil\; rp$	transcription
$= \overline{\overline{q}\rceil\, r}\rceil\;\overline{q}\rceil\; rp$	gen (C2)
$= \overline{\;\;}\rceil\;\overline{q}\rceil\; rp$	gen
$= \overline{\;\;}\rceil$	int (C3).

True.

$2 = \overline{\overline{\overline{\overline{r}\rceil\, p}\rceil\;\overline{pq}\rceil}\rceil\;\overline{p}\rceil}\rceil$	transcription
$= \overline{\overline{\overline{\overline{r}\rceil\, p}\rceil\, p\;\overline{\overline{p}\rceil\, pq}\rceil}\rceil}\rceil$	tra (J2)
$= \overline{\overline{\overline{\overline{\;\;}\rceil\, r}\rceil\, p\;\overline{\overline{\;\;}\rceil\, pq}\rceil}\rceil}\rceil$	gen (C2) (twice)
$= \overline{\overline{\overline{\;\;}\rceil\;\overline{\overline{\;\;}\rceil}\rceil}\rceil}\rceil$	int (C3) (twice)
$=$	pos (J1) (thrice).

False.

$3 = \overline{\overline{\overline{\overline{p}\rceil\, q}\rceil\;\overline{\overline{r}\rceil\, s}\rceil\;\overline{qs}\rceil}\rceil}\rceil$	transcription
$= \overline{\overline{p}\rceil\, q}\rceil\;\overline{\overline{r}\rceil\, s}\rceil\;\overline{qs}\rceil\; pr$	ref (C1)
$= \overline{qs}\rceil\; pr$	occ (C4) (twice).

Contingent.

These calculations, conducted in the primary algebra, are so simple as to be mathematically trivial. That is to say, the moment each of the sentences is written down in the calculus of indications, the answer, to any person familiar with this form, becomes obvious to mere inspection. I have here done the calculations slowly, in very small steps, on the assumption that the reader is not yet familiar with the form.

The consequences of this arithmetical availability are sweeping. All forms of primitive implication become redundant, since both they and their derivations are easily constructed from, or tested by reduction to, a single cross. For example, everything in pp 98–126 of *Principia mathematica* can be rewritten without formal loss in the one symbol

⏋

provided, at this stage, the formalities of calculation and interpretation are implicitly understood, as indeed they are in *Principia*. Allowing some 1500 symbols to the page, this represents a reduction of the mathematical noise-level by a factor of more than 40000.

With such a huge gain in the formal clarity of expressions, the invalidity of a false argument is similarly open to immediate confirmation. We illustrate such an argument below, offered[18] by Maurant as a dilemma.

> If we are to have a sound economy, we must not inflate the currency. But if we are to have an expanding economy, we must inflate the currency. Either we inflate the currency or we do not inflate the currency. Therefore, we shall have neither a sound economy nor an expanding economy.

Let

s stand for *we have a sound economy*

c stand for *we inflate the currency*

e stand for *we have an expanding economy.*

[18] John A Maurant, *Formal logic*, New York, 1963, p 169.

Transcribing it into the primary algebra, we find

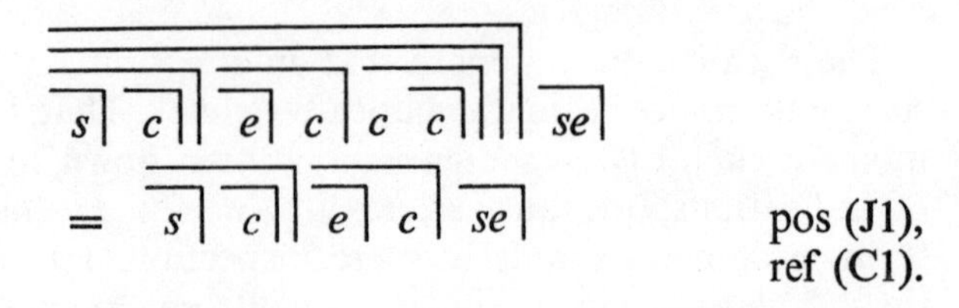

This expression is consequential, but if the fact is not yet apparent, we may use the converse of theorem 16 with an arbitrary variable and constant, say c = ⏋, giving

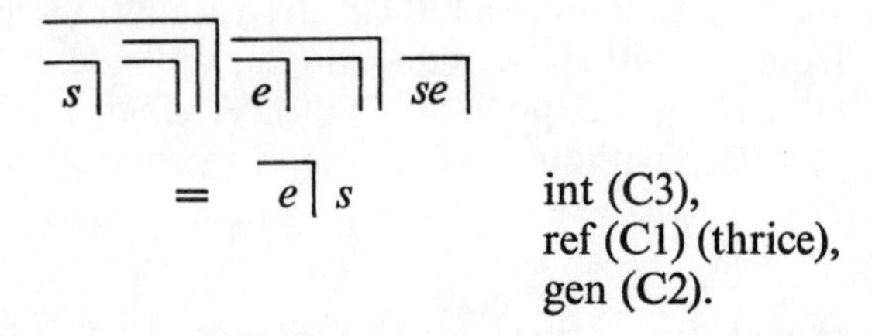

This plainly cannot be reduced, so nor can the original. Thus there is no dilemma. Other characteristics of the argument are also illuminated, especially the utter irrelevance of the premiss 'either c or not c'.

If we stand back for a moment to regard the structure of an implicational logic, such as Whitehead and Russell's, we see that it is fully contained in that of an equivalence logic. The difference is in the kind of step used. In one case expressions are detached at the point of implication, in the other they are detached at the point of equivalence.

If an expression is detached at the point of implication, it of course need not be equivalent to the expression from which it is derived. But if it is a tautology it can be implied only by another tautology, so that, in such cases, the sign of implication can always be replaced by a sign of equivalence. Thus an implicational logic in fact degenerates into an equivalence logic in respect of the class of true statements, with which such logics are most intimately concerned.

The completeness theorem for the primary algebra in the text is what, interpreted in logic, is called a strong completeness theorem, since it includes Post's original[19] weaker theorem. The weaker version merely asserts that all true statements are implied by the true statements initially given as primitive. Since, in the case of true statements, implication is equivalent to equivalence, we see that such a theorem must be included in a theorem which states the completeness of all forms of equivalence, irrespective of whether the statements interpreted from them are true, false, or contingent.

We may turn, now, to consider how the calculus of indications can be applied to the traditional logic of classes. Before doing so, it is of interest to state another hitherto silent (or relatively silent) assumption to the effect that, in the absence of instructions to the contrary, we assume the premisses of an argument to be related by logical conjunction. For example, in transcribing the alleged dilemma above, we first cross the transcription of each individual premiss and then cross the result to give the conjunction, and finally cross all this again for the implication. We have, in fact, habitually come to regard 'and' as the proper interstitial constant. But we could, for example, rephrase both sentential and class logic on the assumption that 'or', instead of 'and', is the constant relating premisses. The reader might like to attempt a proof of this. It is a revealing exercise, especially with respect to the logic of classes, and it is not difficult.

All universal forms of the traditional logic of classes can be accommodated within the logic of sentences, so we will consider these forms first. To accommodate them, we use the pattern in the following key.

for *all a are b* use $(x \in a) \supset (x \in b)$

for *no a is b* use $(x \in a) \supset (x \in \text{not-}b)$

and other forms accordingly. To avoid the use of distinct letters for sentences and classes, we can allow, in the calculating forms, any simple literal variable v to stand for the sentence '$x \in v$', i.e. 'x is a member of the class v'. This will not lead to unintentional confusion, since the sign v, as used to denote the class,

[19] Emil L Post, *Amer. J. Math.*, 43 (1921) 163–85.

does not enter the calculation, which is undertaken with v representing only the truth value of the corresponding sentence.

Taking the form of a syllogism in Barbara, and putting the minor premiss first, as Whitehead and Russell do, we find

if all a are b
and all b are c,
then all a are c

which we can represent by

$$\overline{\overline{\overline{\overline{a}|\,b}|\;\overline{\overline{b}|\,c}|}|}|\;\overline{a}|\,c$$

$$\text{F1} \qquad = \overline{\overline{a}|\,b}|\;\overline{\overline{b}|\,c}|\;\overline{a}|\,c \qquad \text{ref}$$

$$= \overline{\quad}| \qquad \text{gen (thrice), int.}$$

The sentential form is thus seen to be a tautology and the argument thereby valid. In the case of an invalid argument, the algebraic expression will not reduce to a cross, so we have a reliable system for testing the validity of any universal argument in syllogism form. We shall later study a method which will determine the conclusion from the premisses alone. In the present form, as we see, although its validity can be tested, the conclusion, given the premisses alone, can be found only by trial.

Equivalence problems are similarly open to solution in this way.

Example[20]. A club has the following rules.

(*a*) The Financial Committee must be chosen from among the General Committee,
(*b*) No-one shall be a member of the General and Library Committees unless he is also on the Financial Committee,
(*c*) No member of the Library Committee shall be on the Financial Committee.

[20] from B V Bowden, *Faster than thought*, London, 1953, p 36.

Simplify these rules.

Procedure.

for *x is a member of the Financial Committee* write m
for *x is a member of the General Committee* write g
for *x is a member of the Library Committee* write b.

The interstitial constant of a set of rules is usually understood to be conjunction, so we may now transcribe them into the primary algebra as follows.

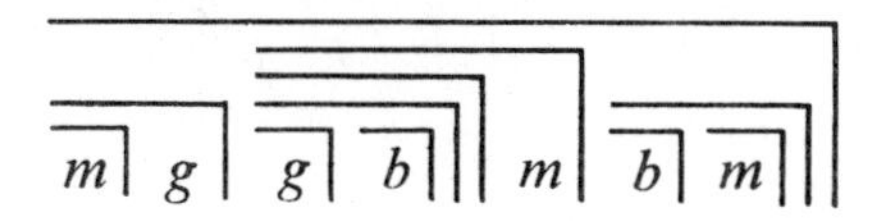

Our aim is to reduce this, if possible, to a simpler conjunctive form which is equivalent to, and may thus be used to replace, the original set of rules.

F2 $\overline{\overline{\overline{m}|\,g}|\;\overline{\overline{g}|\,\overline{b}|\,m}|\;\overline{\overline{b}|\,\overline{m}|}|}|$ ref

$= \overline{\overline{m}|\,\overline{g}|\,b}|\;\overline{m\,\overline{\overline{g}|\,\overline{b}|}|}|$ cro (C9)

$= \overline{\overline{mb}|\,\overline{g}|\,b}|\;\overline{mg}|\;\overline{mb}|$ gen, tra, ref

F3 $= \overline{\overline{g}|\,b}|\;\overline{mg}|\;\overline{mb}|$ gen

$= \overline{\overline{\overline{g}|\,\overline{b}|}|\;\overline{\overline{m}|\,g}|}|\;\overline{mb}|$ cro

$= \overline{\overline{\overline{mb}|\,\overline{g}|\,\overline{b}|}|\;\overline{\overline{mb}|\,\overline{m}|\,g}|}|$ tra

F4 $= \overline{\overline{\overline{g}|\,\overline{b}|}|\;\overline{\overline{m}|\,g}|}|$ occ (twice).

Retranscribing gives the answer

(1) The Financial Committee must be chosen from among the General Committee,
(2) No member of the General Committee shall be on the Library Committee.

We may check this answer by theorem 16. Let $m = \overline{}|$.
Now

$$F2 = \overline{\overline{g}|\; b}|$$

$$F4 = \overline{\overline{\overline{g}|\;\overline{b}|}|\;\overline{g}|}| = \overline{\overline{g}|\; b}|.$$

Let $m =$. Now

$$F2 = \overline{\overline{\overline{g}|\;\overline{b}|}|}|$$

$$F4 = \overline{\overline{\overline{g}|\;\overline{b}|}|}|,$$

so the answer is correct, provided only that we have properly interpreted the problem.

We see that we can, from this answer, obtain an implication (not an equivalence) to the effect that no member of the library committee shall be on the financial committee, since by crossing F4 (for the implication) and reflecting we get

$$\overline{\overline{g}|\;\overline{b}|}|\;\overline{\overline{m}|\; g}|,$$

and now adding our tentative conclusion gives

$$\overline{\overline{g}|\;\overline{b}|}|\;\overline{\overline{m}|\; g}|\;\overline{b}|\;\overline{m}|$$
$$= \overline{}|.$$

The mathematical structure illustrated in this sort of inference suggests the following proposition.

Interpretative theorem 1

If the primary algebra is interpreted so that integral expressions are true, and if each of a number of class-inclusion premisses is sententially transcribed in it, and if variables representing the same sentence at odd and even levels are cancelled, what remains, when retranscribed, is the logical conclusion.

The proof is not difficult and may be left with the reader. The theorem itself, as a short cut to inference, is of considerable power. We may take Lewis Carroll's last sorites to illustrate it.

The problem is to draw the conclusion from the following set of premisses.

(1) The only animals in this house are cats;
(2) Every animal is suitable for a pet, that loves to gaze at the moon;
(3) When I detest an animal, I avoid it;
(4) No animals are carnivorous, unless they prowl at night;
(5) No cat fails to kill mice;
(6) No animals ever take to me, except what are in this house;
(7) Kangaroos are not suitable for pets;
(8) None but carnivora kill mice;
(9) I detest animals that do not take to me;
(10) Animals, that prowl at night, always love to gaze at the moon.

The method employed hitherto to solve such a problem was to work it out by stages, but this can be quite time consuming. Using the theorem above, we simply adopt a distinct variable for each distinct (but not complementary) set, transcribe, cancel, and arrive at the answer practically instantaneously. Let us, then, proceed to adopt

h for *house, in this*

c for *cat*

p for *pet, suitable for*

d for *detested by me*

a for *avoided by me*

m for *moon, love to gaze at*

v for *carnivorous*

n for *night, prowl at*

k for *kill mice*

t for *take to me*

r for *kangaroo.*

We see from the principle of relevance that we do not need to adopt a variable for the set of animals. We now proceed to the transcription and cancellation

$$\overline{\cancel{h}}\rceil\ \cancel{c}\ \overline{\cancel{m}}\rceil\ \cancel{p}\ \overline{\cancel{d}}\rceil\ a\ \overline{\cancel{p}}\rceil\ \cancel{v}\ \overline{\cancel{c}}\rceil\ \cancel{k}\ \overline{\cancel{t}}\rceil\ \cancel{k}\ \overline{r}\rceil\ \overline{\cancel{p}}\rceil\ \overline{\cancel{k}}\rceil\ \cancel{v}\ \cancel{t}\ \cancel{d}\ \overline{\cancel{n}}\rceil\ \cancel{m}$$

which reveals $\overline{r}\rceil\ a$. Therefore, all kangaroos are avoided by me.

So far we have considered how the calculus of indications, in the form of the primary algebra, may be used to clarify and simplify problems in sentential logic, and also those of universal, or non-existential, import in class logic or set theory. We shall turn now to consider its extension, in class logic, to problems of existential, or particular, import.

We resolved the question of how to represent a universal statement such as

all *a* are *b*

by translating it into an equivalent complex in the sentential calculus. The question we must now seek to answer is, can an existential statement, such as

some *a* are *b*,

be similarly translated?

We first note that, to contradict the general assertion that all *a* are *b*, it is sufficient to find some *a* that are not *b*. We may note by the way that the statement no *a* is *b*

does not contradict

all *a* are *b*

since, in case *a* is non-existent, both assertions are true.

Transcribing according to the principles already adopted, we take

some *a* are not *b*

to say

not all *a* are *b*

and so represent it by

$$\overline{\overline{a|}\ b\,}| \; .$$

Similarly we represent

some *a* are *b*

by

$$\overline{\overline{a|}\ \overline{b|}}| \; .$$

To see how this works out, we transcribe another syllogism, this time of existential import. Thus

all *a* are *b*
some *a* are *c*
∴ some *b* are *c*

becomes

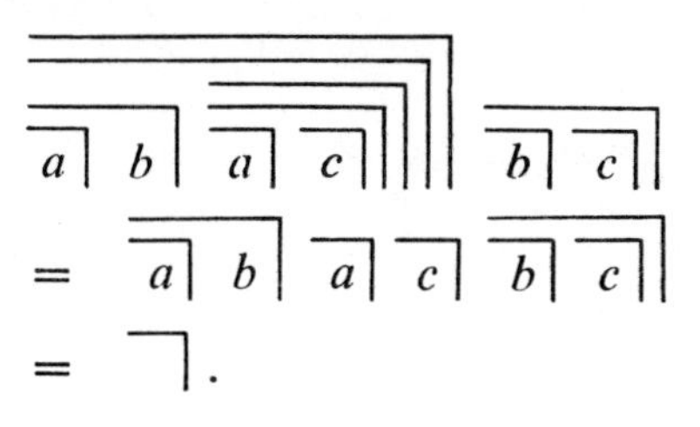

Since we can see otherwise that this syllogism is valid, it appears to be properly represented. But using the same rules we can represent

some *a* are *b*
some *b* are *c*
∴ some *a* are *c*,

which we know to be invalid, by

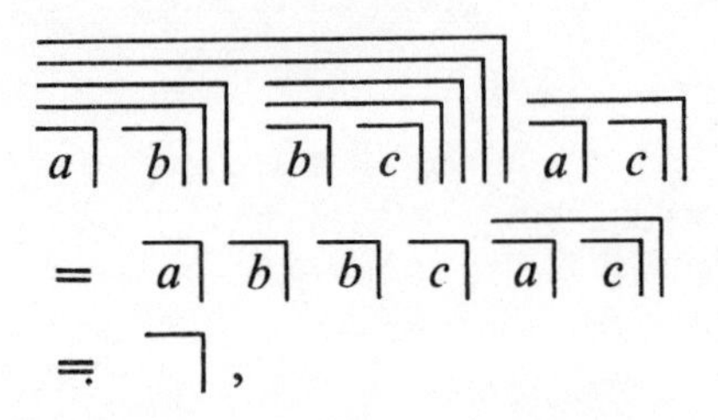

in which it appears to be valid. How do we resolve this seeming contradiction?

Let us be clear of one thing. The question is answered (implicitly, since it is not usually asked) in the textbooks as it was originally answered by Aristotle, by giving a more or less complicated set of rules which disallow this inference. But a set of rules to say *that* one must not do something is not an explanation of *why* one must not, and nor does the fact that, if we allow the inference, it *may* mislead us to an improper conclusion, meet with the high degree of understanding required of all explanatory* accounts in this book. We have found an area in which an apparently impeccable interpretative procedure has suddenly let us down, and all rules which say we must *therefore* avoid this area, however well they may work in practice, have an unsatisfactorily *ad hoc* flavour.

* To *explain,* literally to lay *out* in a *plane* where particulars can be readily seen. Thus to *place* or *plan* in *flat* land, sacrificing other dimensions for the sake of appearance. Thus to *expound* or *put out* at the cost of ignoring the *reality* or *richness* of what is so put out. Thus to take a view away from its *prime reality* or *royalty*, or to gain knowledge and lose the kingdom.

It is no use, either, appealing to graphical forms such as Venn diagrams, since these, in common with other graphs, offer a picturesque realization which is peripheral, not central, to the question. To answer it, we must find an altogether subtler approach.

We begin with the observation that statements about the universe of discourse, e.g.

if there is an a in it,
then this a is also a b in it,

assert no claim to the *existence* of anything in it, although they may be taken, at a different level, to claim the existence of the universe which has these conditional properties. But to deny such a statement, we claim that

there *exists* at least one a in it which is not a b.

Now the distinction between existing and not existing is not applied like the distinction between true and not true. If a statement s is true, then its complementary statement $\sim s$ is false. But if a thing t exists, then its complementary thing not-t is not necessarily non-existent. In the universe of England, the complement of London is the country. Both, at the time of writing, apparently exist. Thus no existence *follows* from another existence, so that from a statement, or a list of statements, asserting existence only, no proper conclusion can be drawn.

So far we are still at the periphery. That is to say, we are still examining the form of the interpretation, without finding exactly how and where it breaks faith with the mathematics.

In relating the mathematics and the interpretation, we found forms such as

$$\overline{a\,|}\; b$$

which say nothing, in their interpretation, about existence, neither asserting nor denying it. But such forms, when crossed,

$$\overline{\overline{a\,|}\; b\,|}\,,$$

now do say something about existence, at least in the interpretation we have allowed them.

The expression $\overline{a|}\; b$ is universal because it limits the shape of the universe so that there is no space in it for an *a* which is not a *b*. At least, that is how we take it. But we *could* (although we don't) take it to mean, simply, that in this universe there just happens not to exist an *a* which is a *b*, although there is the space, if we want it, to hold one. In other words, we could (although we don't) interpret it existentially.

Similarly we *could* interpret the expression $\overline{\overline{a|}\; b\,|}$ universally. For the statement that some *a* are not *b*, although *sufficient* to contradict the statement that all *a* are *b*, is not *necessary*. An alternative way of contradicting it would be simply to deny that the universe is of such a form as to demand of any *a* that it shall be also a *b*, without actually requiring the existence of an *a* to prove it.

In this alternative we have a means of confining all interpretations to a non-existential import. Let us see how it works out in the case of the invalid syllogism. We should now write

some *a* are *b*
some *b* are *c*
∴ some *a* are *c*

in the form

it is not the case that no *a* is *b*
it is not the case that no *b* is *c*
∴ it is not the case that no *a* is *c*,

making explicit the requirement than no statement is to be taken existentially.

Even so, at first sight, we are not entirely out of trouble. For although, from such a description, the universe appears compelled to reserve space for *a*'s which might be *b*'s and for *b*'s which might be *c*'s, it does not appear *compelled* (as, by the implication, it should be) to reserve any space for *a*'s which might be *c*'s.

But a universe without such space would contain at most six different departments, since it would be missing a department for *a*'s which are also *b*'s which are also *c*'s, and for *a*'s which are also not-*b*'s which are also *c*'s. Now there is a well known theorem, a proof of which was published [14, p 309] by Huntington in 1904, according to which *the number of elements in every finite logical field must be* 2^m (*m an integer* >0). Thus an algebra suitable for such a logical field cannot, without further constraint, represent a form in which the number of elements is not a natural integral power of 2. Such a constraint, when required, is normally imposed through the premisses. That is to say, if any of the possible 2^m spaces is required to be absent from the universe, it must be positively (i.e. referentially) excluded. None of the possible eight spaces is excluded by the premisses of the syllogism above, and so all eight must be presumed to exist, or the mathematical form cannot be properly interpreted. And if they exist, the conclusion follows.

Another, and perhaps easier, way to see *in what sense* the traditionally invalid syllogism above is valid, is to return to our original method of interpretation. Using standard sentential constants it becomes

$$\sim((x \in a) \supset (x \in \text{not-}b))\,.\sim(\sim(x \in \text{not-}b) \supset (x \in \text{not-}c))$$
$$\supset \sim((x \in a) \supset (x \in \text{not-}c))$$

and is of course, in this form, true.

Let there be no mistake, we do not assert, by this, that the syllogism taken, isolated, within the ordinary meaning of 'some *a* are *b*, some *b* . . . etc' is anything other than invalid. It is just that, in trying to place it in a deeper mathematical foundation, we come across (or up against) the inconsequential relation, apparent in ordinary speech, between a form and its content, occasioned by the partly accidental fact that the existence of a particular content can serve to negate a general form.

It remains for us to extricate ourselves, as elegantly as we can, from the unintentional confusion which follows in the train of such a state of affairs: or alternatively, if we have so extricated ourselves, to devise the most peaceful set of rules by which the possibility of such confusion can be laid to rest. The rules which, by tradition, are enlisted to serve this purpose are too numerous for what is a basically simple ambiguity, and they may surely be reduced.

Such a reduction, as we have seen, will be mathematically powerful if it can be taken to a point of degeneration. In this case the ideal degeneration would be at a place where the two kinds of denial, universal or existential, of a universal proposition amount to the same thing. At such a point we could use the calculus freely, without fear of its letting us down.

We have observed that as long as inferences or equations in class logic are universally interpreted, the primary algebra can be freely used to determine them. In other words, the sentential form into which we placed universal statements about classes or sets can be seen to accommodate then exactly, without formal loss or gain. It is the denials of such statements, when we wish to interpret them existentially, that present the difficulty, which arises evidently from a formal gain, since we find a need to constrain the calculus in this respect, rather than to relax it.

Let us return, for a moment, to examine our procedure for solving Bowden's problem about the club rules. In the algebraic path to its solution we find an expression

$$\overline{\overline{g|}\; b\,|}\;\; \overline{mg|}\;\; \overline{mb|}\;,$$

marked F3. Taken existentially, it would mean

either some g are not b
or some things are neither m nor g
or some things are neither m nor b.

But in fact the whole argument depends on not taking F3, or any other intermediate expression, this way. Algebraically, of course, it doesn't matter, we have no choice, and arrive at the answer willy-nilly. It is only on retracing the path by which we got there, and stopping on the way to look at the pitfalls, that we see the alarming prospect of the interpretative dangers which it effectively by-passed.

The first rule which suggests itself, therefore, is never to make an existential interpretation unless the argument demands it. No such demand is evident in Bowden's problem, and so, in solving it, we can effectively avoid existence, and thereby avoid the pitfalls it brings in its train. The question which then frames itself is how far we can take this avoidance, or, considered in reverse, in what circumstances, and at what place, during the course of solving a problem, do we ever need to make an existential interpretation?

The answer is none. Existential interpretations, where they are necessary at all, can be confined to entering and leaving the problem, and need never occur in the course of solving it.

To see how this comes about, we may return to the syllogism in Barbara, taken in the form

F1
$$\overline{\overline{a\,}|\;b\,}|\;\;\overline{\overline{b\,}|\;c\,}|\;\;\overline{a\,}|\;c\,.$$

Since the order of each of the three complexes in F1 is irrelevant to the meaning of the whole expression, we may transpose it to find

F1′
$$\overline{\overline{a\,}|\;b\,}|\;\;\overline{a\,}|\;c\;\;\overline{\overline{b\,}|\;c\,}|$$

$$=\;\overline{\overline{\overline{\overline{a\,}|\;b\,}|\;\;\overline{\overline{\overline{a\,}|\;c\,}|\,}|\,}|\,}|\;\;\overline{\overline{b\,}|\;c\,}|\,,$$

which can be retranscribed

all a are b
some a are not c
∴ some b are not c.

Transposing it yet again, we find

F1″
$$\overline{a\,}|\;c\;\;\overline{\overline{b\,}|\;c\,}|\;\;\overline{\overline{a\,}|\;b\,}|$$

which will give

some *a* are not *c*
all *b* are *c*
∴ some *a* are not *b*.

So we see that the representative form of a syllogism in Barbara *is also* the representative form (remembering that we have in each case put what is called the minor premiss first) of syllogisms in Bocardo and Baroco. The three syllogisms above, being effectively reducible to the same mathematical expression, must therefore represent, at this level, an *identical* form of argument.

This is both interesting and fascinating. It is interesting because, from it, we shall be able to obtain a much simplified rule-structure for existential arguments, and fascinating because of the light it sheds on what we are doing when we argue from existence. We may note in passing, as Prior reminds[21] us, that glimpses of the path to this identity are apparent in the work of Aristotle, who refers to a form lately more fully described[22] by Ladd-Franklin, in which what she calls an antilogism condenses three syllogisms. Here we elucidate a further stage, in which the three-in-one nature of the syllogism is evident from its transcription alone, without recourse to an image or antilogism.

From the conversion (or converse) of what we have just recounted, we observe the following proposition.

Interpretative theorem 2

An existential inference is valid only in as far as its algebraic structure can be seen as a universal inference.

For example, each of the existential arguments transcribed from F1′ and F1″ is valid because of the validity of the universal argument transcribed from F1.

[21] A N Prior, *Formal logic*, 2nd edition, Oxford, 1964, p 113.
[22] C F Ladd-Franklin, *Mind*, 37 (1928) 532–4.

This single rule takes care of all the separate rules for syllogisms, their parts, and their extensions. It even includes the provision that there shall be not more than one particular premiss, for with more than one, no representation as a universal argument is possible.

We have here found the degeneration we were seeking, at the place where the existential condenses with the universal. This degeneration, like the one undertaken earlier for the sentential calculus, is a release from the bond of the particular, and through it we see the whole syllogistic structure in the one prototype

$$\overline{\overline{a|}\; b|}\;\overline{\overline{b|}\; c|}\;\overline{a|}\; c\ .$$

In this prototype, not only can we transpose each complex, we can also independently cross each literal variable, finding, by a combination of these means, a set of 24 distinguishable valid arguments. Formally there is no difference between them. If we distinguish any, we should distinguish all. In fact not all twenty-four are distinguished in logic, which arrives somewhat arbitrarily at the number fifteen.

Thus, by realizing a condensation, we no longer need to remember, for syllogisms or related arguments, the wearisome rules of their construction and validity. All these are now subsumed in, and can be reconstructed from, the simple basic form and interpretation to which we have here reduced them.

We may return, for a moment, to reconsider the sorites, which is the general form under which the syllogism is the primary member. In the light of the degeneration undertaken above, we see that the method we developed for revealing a conclusion by cancellation applies equally whether the argument is universal or existential. For a universal sorites we have

$$\overline{a|}\; b,\ \overline{b|}\; c, \ldots,\ \overline{p|}\; q,$$

$$\therefore \qquad \overline{a|}\; q.$$

To convert it into an existential one we simply negate the

conclusion and transpose it with one of the premisses which, itself negated, becomes the new conclusion. So from

$$\overline{a|}\ b,\quad \overline{b|}\ c,\quad \overline{c|}\ d,$$

$$\therefore\qquad \overline{a|}\ d,$$

we find, for example, the set of premisses

$$\overline{a|}\ b,\quad \overline{b|}\ c,\quad \overline{\overline{a|}\ d\,|}$$

from which, as before, the conclusion can be revealed by cancellation,

$$\overline{\not a|}\ \not b\ \ \overline{\not b|}\ \ c\ \ \overline{\overline{\not a|}\ d\,|}\ .$$

All we need to remember is that it will now be existential, and so should in this case be written

$$\overline{\overline{c|}\ d\,|}\ .$$

We leave the account here, where the interested reader will be able to continue it at his pleasure. The problems solved so far, and the questions answered, are simple ones, although the calculus is, in practice, successfully applied[23] to the solution of problems of great complexity. So much, at the primitive level, is commonly overlooked, and what is seen is normally recounted in a fashion so fragmentary as to be hardly coherent. The very act of dwelling for a while with even a simple form can evidently tax the whole of one's powers, so that to leave the simple forms before one is properly familiar with them can result in many unrewarding, or largely unrewarding, mathematical excursions.

To be concerned, as we have been, with what can be found, if we seek it, at a level of extreme simplicity, is in the way of

[23] Cf George Spencer-Brown, British Patient Specifications 1006018 and 1006019 (1965).

being beyond the elementary, but beyond on the side of simplicity, not complexity. This does not, of itself, make what is written here easy to follow, but if the reader is ready to build with charity upon its insufficiencies he may find in it enough reward to do justice to his and my labours.

INDEX OF REFERENCES

Note. In context, a page reference is confined to what is of particular interest to the discussion. In this index it is expanded to include the whole work.

[1] George Boole, *The mathematical analysis of logic*, Cambridge, 1847. xi

[2] **Alfred North Whitehead and Bertrand Russell, *Principia mathematica*, Vol. I, 2nd edition, Cambridge, 1927.** xii

[3] **Henry Maurice Sheffer, *Trans. Amer. Math. Soc.*, 14 (1913) 481–8.** xii

[4] Ludwig Wittgenstein, *Tractatus logico-philosophicus*, London, 1922. xiv

[5] Kurt Gödel, *Monatshefte für Mathematik und Physik*, 38 (1931) 173–98. xv

[6] Alonzo Church, *J. Symbolic Logic*, 1 (1936) 40–1, 101–2. xv

[7] W V Quine, *J. Symbolic Logic*, 3 (1938) 37–40. xv

[8] Abraham A Fraenkel and Yehoshua Bar-Hillel, *Foundations of set theory*, Amsterdam, 1958. xvii

[9] P B Medawar, Is the Scientific Paper a Fraud, *The Listener*, 12 September 1963, pp 377–8. xviii

[10] R D Laing, *The politics of experience and the bird of paradise*, London, 1967. xviii

[11] Edward V Huntington, *Trans. Amer. Math. Soc.*, 35 (1933) 274–304. 88

[12] Alfred North Whitehead, *A treatise on universal algebra*, Vol. I, Cambridge, 1898. 90

[13] Charles Sanders Peirce, *Collected papers*, Vol. IV, Cambridge, Massachusetts, 1933. 90

[14] *ΠΡΟΚΛΟΥ ΔΙΑΔΟΧΟΥ ΣΤΟΙΧΕΙΩΣΙΣ ΘΕΟΛΟΓΙΚΗ* with a translation by E R Dodds, 2nd edition, Oxford, 1963. 90

[15] Edward V Huntington, *Trans. Amer. Math. Soc.*, 5 (1904) 288–309. 96

[16] George Boole, *An investigation of the laws of thought*, Cambridge, 1854. 97

[17] E Stamm, *Monatshefte für Mathematik und Physik*, 22 (1911) 137–49. 111

[18] John A Maurant, *Formal logic*, New York, 1963. 117

[19] Emil L Post, *Amer. J. Math.*, 43 (1921) 163–85. 119

[20] B V Bowden, *Faster than thought*, London, 1953. 120

[21] A N Prior, *Formal logic*, 2nd edition, Oxford, 1964. 132
[22] C F Ladd-Franklin, *Mind*, 37 (1928) 532–4. 132
[23] George Spencer-Brown, British Patent Specifications 1006018 and 1006019 (1965). 134

INDEX OF FORMS

Note. A theorem marked with an asterisk has a true converse.

Principle of relevance 43

If a property is common to every indication it need not be indicated

Principle of transmission 48

With regard to an oscillation in the value of a variable, the space outside the variable is either transparent or opaque

Rule of demonstration 54

A demonstration rests in a finite number of steps

ARITHMETICAL INITIALS

I1	$\overline{\;\;}|\;\overline{\;\;}| \;=\; \overline{\;\;}|$	number	12
I2	$\overline{\overline{\;\;}|}| \;=$	order	12

ALGEBRAIC INITIALS

J1	$\overline{\overline{p}|\;p|}| \;=$	pos	28
J2	$\overline{\overline{pr}|\;\overline{qr}|}| \;=\; \overline{\overline{p}|\;\overline{q}|}|\;r$	tra	28

THEOREMS

representative

*T1 The form of any finite cardinal number of crosses can be taken as the form of an expression 12

T2 If any space pervades an empty cross, the value indicated in the space is the marked state 13

T3 The simplification of an expression is unique 14

T4 The value of any expression constructed by taking steps from a given simple expression is distinct from the value of any expression constructed by taking steps from a different simple expression 18

procedural

T5 Identical expressions express the same value 20

*T6 Expressions of the same value can be identified 20

*T7 Expressions equivalent to an identical expression are equivalent to one another 21

CONSEQUENCES

C1	$\overline{\overline{a}\vert}\vert = a$	ref	28
C2	$\overline{ab}\vert\ b = \overline{a}\vert\ b$	gen	32
C3	$\overline{\ }\vert\ a = \overline{\ }\vert$	int	32
C4	$\overline{\overline{a}\vert\ b}\vert\ a = a$	occ	33
C5	$aa = a$	ite	33
C6	$\overline{\overline{a}\vert\ \overline{b}\vert}\vert\ \overline{\overline{a}\vert\ b}\vert = a$	ext	33
C7	$\overline{\overline{\overline{a}\vert\ b}\vert\ c}\vert = \overline{ac}\vert\ \overline{\overline{b}\vert\ c}\vert$	ech	34
C8	$\overline{\overline{a}\vert\ \overline{br}\vert\ \overline{cr}\vert}\vert = \overline{\overline{a}\vert\ \overline{b}\vert\ \overline{c}\vert}\vert\ \overline{\overline{a}\vert\ \overline{r}\vert}\vert$	mod	34
C9	$\overline{\overline{\overline{b}\vert\ \overline{r}\vert}\vert\ \overline{\overline{a}\vert\ \overline{r}\vert}\vert\ \overline{\overline{x}\vert\ r}\vert\ \overline{\overline{y}\vert\ r}\vert}\vert$		
	$= \overline{\overline{r}\vert\ ab}\vert\ \overline{rxy}\vert$	cro	35

G SPENCER BROWN is by training a man of science. He studied medicine at the London Hospital Medical College and psychology at Cambridge University, and later worked with Wittgenstein and Russell in philosophy and mathematics. He held academic positions at both Oxford and London Universities, and also writes non-mathematical books under the name James Keys.

Reader's Notes